Free Will and Modern Science

Free Will and
Modern Science

Edited by

Richard Swinburne

Published for THE BRITISH ACADEMY
by OXFORD UNIVERSITY PRESS

Oxford University Press, Great Clarendon Street, Oxford OX2 6DP

Oxford New York
Auckland Cape Town Dar es Salaam Hong Kong Karachi
Kuala Lumpur Madrid Melbourne Mexico City Nairobi
New Delhi Shanghai Taipei Toronto

With offices in
Argentina Austria Brazil Chile Czech Republic France Greece
Guatemala Hungary Italy Japan Poland Portugal Singapore
South Korea Switzerland Thailand Turkey Ukraine Vietnam

Published in the United States By Oxford University Press Inc., New York

British Library Cataloguing in Publication Data
Data available

Library of Congress Cataloging in Publication Data
Data available

Typeset by Keystroke, Station Road, Codsall, Wolverhampton
Printed in Great Britain on acid-free paper by
CPI Group (UK) Ltd, Croydon, CR0 4YY

ISBN 978-0-19-726489-8

Contents

Foreword: questions of freedom   vii
PETER SIMONS

List of figures   xvi

List of contributors   xvii

Introduction: plan of the volume   1
RICHARD SWINBURNE

1 Does brain science change our view of free will?   7
   PATRICK HAGGARD

2 Libet and the case for free will scepticism   25
   TIM BAYNE

3 Physicalism and the determination of action   47
   FRANK JACKSON

4 Dualism and the determination of action   63
   RICHARD SWINBURNE

5 On determinacy or its absence in the brain   84
   HARALD ATMANSPACHER AND STEFAN ROTTER

6 Gödel's incompleteness theorems, free will and
   mathematical thought   102
   SOLOMON FEFERMAN

7 Feferman on Gödel and free will: a response to Chapter 6   123
   J.R. LUCAS

Contents

8 The impossibility of ultimate responsibility? 126
GALEN STRAWSON

9 Moral responsibility and the concept of agency 141
HELEN STEWARD

10 Substance dualism and its rationale 158
HOWARD ROBINSON

11 What kind of responsibility must criminal law presuppose? 178
R.A. DUFF

    Guide to background reading 200
    Index of names 203

Foreword

Questions of freedom

PETER SIMONS

The problem of free will is one of the oldest in all philosophy and has long resisted solution. It has been discussed and disputed in all epochs of the subject, from the ancient Greeks, through the medieval and early modern period, to today. There is, as this volume graphically demonstrates, still no informed consensus as to its solution. About the only thing that all parties to the dispute agree on is its importance. Nearly all people are perfectly convinced, whether instinctively or upon reflection, that normally endowed, rational, adult human beings have a measure of choice in what we do, and that some of the time, when we act, we do so of our own free will. What this means exactly is part of the problem, but the widespread conviction that we do have freedom in some of our choices and actions is a given. It is anchored in our attitudes, assumptions and institutions.

The traditional term 'free will' (or 'freewill') goes back to the idea that there is a special faculty or capacity of the soul, mind or person, the will, which we may exercise or operate freely on occasion. But the idea is describable and familiar independently of this assumption about human psychology. In many ways 'free action' or 'free choice' are less loaded descriptions. We shall not be choosy about terminology here: it is the phenomenon that is important.

Any number of examples may be instanced as putative cases of the operation or exercise of free will. Historic events of momentous importance may be set in train by the decision of a ruler or other person with power: Napoleon's decision to invade Russia, or President Truman's decision to drop atomic bombs on Japan in 1945, for example. Most of us are familiar with choices of less world-historic moment: what wine to have with a meal, where to go for a holiday, what career to pursue, whether to take up a job offer. We assume without question that, despite influences and inclinations, what we do in such cases is ultimately 'up to us'. My choices, great or small, are up to me, it is I who takes the decision, makes the attempt, executes the action and assumes responsibility for what I do. Everyone is so familiar with such cases from his or her own life that it needs little elaboration. That is the datum: call it the *freedom assumption*.

1. Determinism and the clash with freedom

The problem, or problems, of free will arise when the freedom assumption comes into contact, and conflict, with other general propositions from common-sense and systematic, scientific thought. The most pervasive conflict is with forms of determinism, which are views according to which whatever happens in the world, an event, is to be accounted for or explained by tracing the causes of that event. At all stages of systematic thinking, people have looked for accounts of events which render what happens intelligible in the light of what went before. Historians account for Truman's decision to use the atomic bomb in terms of the wish to avoid massive casualties in attempting to invade the Japanese homeland, in the knowledge of the high casualties sustained in Okinawa and elsewhere. They also cite Truman's attitudes, the advice and other factors influencing him, the institutional momentum of the Manhattan Project, and so on. We know that the explanation of such historic actions is complicated and difficult. In the natural sciences, explanation of events is more straight-forward. We account for the symptoms of an illness in terms of its underlying causes, for example the symptoms of a cold are caused by the presence of large numbers of rhinoviruses in the respiratory tract, the fading of a fabric by the chemical changes to dyes caused by ultra-violet radiation in sunlight, and so on.

Since the beginnings of scientific thought over two thousand years ago, theorists have looked to find accounts of what happens in terms of the exceptionless conformity of sequences of events to a relatively small number of fixed patterns. The aim of science, it seems, is to describe these patterns so that all events constitute instances of them in combination. The regularities thus uncovered are the laws of nature. Whether these laws are laid down by a divine lawgiver or simply the unbroken habits of natural processes is here not so important. What matters is that all events, including human actions, come about in conformity with the laws of nature.

As science developed, this form of explanation became less speculative and assumed greater clarity, precision, and universality of scope. In Newton's mechanics, the motions of heavenly and earthly bodies alike are taken as following a small number of simple laws, and a wide variety of cases is describable using the idiom of geometry. The apogee of this development was Pierre-Simon Laplace's *Mécanique céleste* of 1799–1825. The deterministic assumption behind this cosmology Laplace enunciated in his *Essay on Probability* of 1795:

> We ought then to regard the present state of the universe as the effect of its anterior state and the cause of the one which is to follow. Given for one instant an intelligence which could comprehend all the forces by which nature is animated and the respective situation of the beings who compose it – an intelligence sufficiently vast to submit this data to analysis – it would embrace in the same formula the movements of the greatest bodies of the universe and those of the lightest atom; for it, nothing would be uncertain and the future, as the past, would be present in its eyes.

This is a clear statement of universal causal determinism: given the total state of the universe at a particular time, the laws of nature dictate its state at all other times, past as well as future.

Since human decisions and actions are events taking place in the world, it would seem that these too are subject to universal causal determinism. If this is right, then our actions cannot be 'up to us'. My decision to go for a coffee and my ensuing action are events which, given the state of the universe at any time, were bound to happen and in principle (though not in practice) predictable from any time in the past, including all the long eons before I was born. It seems that although I seem to have a choice, to be able on my own to determine what I do, in fact I do not have such a choice. All my actions, all the things that happen to me, including those events that I take to be my free choosings, are fixed and were fixed long ago. History runs on rails, and freedom is an illusion.

2. Accepting the clash – hard determinism versus libertarianism

The opposition between the freedom assumption and determinism is asymmetric, in that one proposition, the freedom assumption, is a widely accepted and intuitive view based on experience, while the scientific hypothesis is both more speculative and more remote in its strict version from experience. Nevertheless many philosophers, scientists and theologians have felt the strength of the opposition. The view that freedom and determinism cannot both be true is called *incompatibilism*. Incompatibilists therefore face a theoretical choice: whether to reject the freedom assumption, or the deterministic hypothesis, or indeed both. Accepting determinism then entails denying freedom: this view is what William James called 'hard determinism'. Several philosophers have been hard determinists: some Greek Stoics, and later Spinoza, as well as some Muslim doctors and Protestant theologians such as Calvin and Luther, who ascribe ultimate freedom to determine events to God alone, and the view has also been supported by scientists like Clifford and Huxley. The consequences of accepting hard determinism are highly revisionary, to say the least, so this has not been a popular option.

The more widespread attitude among incompatibilists has been to deny determinism. In a sense this is easily done: universal determinism states that all events are causally determined, or determined in some other way (we will not consider all the variants here). It takes only a single event which is not determined to make determinism false. Epicurus and ancient Epicureans took this way out, proposing that atoms sometimes spontaneously and without reason or cause swerve in their paths. Indeterminism is indeed the majority view among modern physicists, embodied in standard interpretations of quantum theory. Does indeterminism thus make room for freedom?

The answer is: not necessarily. First, it might be that all events which are not fully determined happen in places that are far removed from human beings, while events in and around us are fully determined. So the truth of indeterminism is insufficient to guarantee freedom. Secondly, even if, as quantum theory would seem to imply, the universe including ourselves and our parts is replete with indeterministic events, these might be at a level of scale or granularity which has no bearing on our decisions and actions, being at the subatomic and not the cellular level. Events which at a microscale are random and undetermined may be 'smoothed out' by a larger scale when large numbers of them are considered. For example, in the classic two-slit experiment, individual photons passing through the apparatus may impinge at any point on a target screen or photographic plate, but when large numbers of them do so the result is a smoothed out classical wave pattern of the sort described by classical physics. It might be that the indeterminism of quantum events is all smoothed out by the level at which we consider human beings and their physiological processes, such as neuron-firings, which would render events at this level subject to the determinism of classical mechanics. Opinions differ as to whether this is in fact what happens in the quantum case, so this is an open question. It should also be mentioned in passing that there are deterministic inter-pretations of quantum mechanics, so a simple appeal to the advance of modern physics does not settle the question as to the truth or falsity of determinism.

Nevertheless, whether for quantum-theoretical or other reasons, merely having exceptions to determinism does not generally make for a satis-factory underpinning to the freedom assumption. If quantum events, like Epicurean atomic swerves, take place spontaneously at the micro-level and are implicated in our conscious decisions, how can they be said to con-stitute or contribute to freedom? No one would say that the radioactive nucleus that spontaneously splits at a certain moment is exercising its freedom, so why should a battery of such random events in the brain make a human being free? They would appear to be just as buffeted by events as a leaf in the wind. What is missing from the picture is the idea that a free agent is, at least up to a degree, in control, that they themselves determine, at least in part, what happens. This is what is meant by saying that some things are 'up to us'. It is not merely that they might turn out this way or that, but that it is the agent who determines whether it is this way or that, and not some random event or aggregate of events.

Among incompatibilists who accept freedom, who are known as *libertarians*, it is therefore common to suppose that agents intervene in the course of events in a way which differs from the chance or randomness of quantum physics, but gives them a form of determination not subject to the laws of causation applying among other natural events. Some events have their ultimate causal origin in the agent, or in the agent's free decisions. The idea of a form of causation which goes not from event to event but from agent to event, so-called *agent causation*, is not new. It is found in Aristotle and Locke, and also informs those kinds of theism where God is outside space (and perhaps also time) but nevertheless intervenes in temporal affairs. Agent causation has seen something of a revival in recent years, principally as a response to the problem of free will. The nature of the agent at the origin of agent causation is disputed and varies according to theory. Some agents are conceived as natural, spatiotemporal beings, organisms; other views take the agent outside of space as something like a traditional soul. Both views are represented in this volume.

3. Denying the clash – compatibilism

Many philosophers deny and have denied that freedom and determinism are in conflict. They are called compatibilists. Some Stoics may have held this view, and it is quite popular among early modern British philosophers: Hobbes, Hume and Mill are all compatibilists. The standard way compatibilists justify their view is to claim that what we mean by 'free' and similar expressions has nothing to do with being somehow outside the standard causal network and everything to do with whether what we do is caused by us or forced on us by agencies – other people, physical circumstance, inner compulsion – beyond our control. So, to consider a classical example of Hume's, a prisoner confined and in chains is not free to stroll in the park or down to the pub for a drink, but is forced to remain in his cell. Likewise a person falling from a ladder is not free to fly back up to their perch, a drug addict injecting heroin is not free to refrain from doing so, an epileptic jerking about during a seizure is not free to sit still. What is happening to them is not under their control. When it is, that is when they are free, and such free actions are free notwithstanding their being part of the causal network of events as explicable by science.

The compatibilist thus seeks to have the assurance of freedom, avoiding the extreme revisionary consequences of denying it, while accepting the universality of causal determination as valued in natural science. For much of the twentieth century, compatibilism was the dominant opinion among philosophers, but not all are satisfied that it solves the problem, and it is less popular than it was. The principal problem with compatibilism is that the kind of absence of coercion or compulsion that compatibilists call 'freedom', while not irrelevant to free will, is felt to fall significantly short of the positive voluntary selection of different physically possible futures that the freedom assumption is generally thought to entail. The description of freedom as the absence of various constraints makes the concept of freedom too ad hoc and subject to whim and variation: if someone comes up with a new way to constrain someone – say by remote manipulation of brain electricity – then this will get added into the conditions which proscribe freedom. For just such reasons William James, who called compatibilism 'weak determinism', considered it a 'quagmire of evasion'.

On the other hand the compatibilist will point to such elasticity as precisely the virtue of her approach. If in the history of the world no one had hitherto manipulated brainwaves of others from their mobile phone, then this new development alters what we consider freedom to consist in. Perhaps it does not require a lexical redefinition of the word 'freedom' but it constitutes a new factor in the openly negative account of what freedom itself consists in. Compatibilists can also point out that our understanding of the extent of an individual's moral responsibility shifts as we discover new facts about what aspects of their behaviour and their character people are able to control. Whereas soldiers were shot for cowardice or desertion in the First World War, similar cases were treated as meriting not execution but hospitalization and treatment in the Second World War. As we discover more about human beings through physiology and medical science we find out more about the causes and constraints of their behaviour, including their apparently free actions.

4. Freedom and brain science

The scientific investigation of the way the brain functions and how it interacts with overt behaviour and consciousness is barely a century and a half old, and much of the advance in our understanding is much more recent, as new investigative tools such as neuroimaging have become available. It is fair to say that there is much more to discover than we know already. Only with the availability of such tools has it become possible to delve empirically into the connections between putative cases of free choice and the brain. So to a fair extent, questions that had to be answered provisionally, appealing to introspection, phenomenology and verbal report, can be looked at from an ethically admissible, non-invasive neurophysiological point of view. Such is the case with the now famous experiments of Benjamin Libet, discussed in this volume, which may seem to show that, as fatalists had long held, freedom is an illusion. Without entering into details, what Libet's experiments appear to show and have been taken by some to show, is that when a human being is put in an experimental situation of being required to make a choice, the processes initiating the action of doing one of the alternatives occur *before* the subject's conscious experience of making the choice. In other words, physically the action is already under way, so the awareness of choosing is not causally effective in deciding which action it is to be. Unless causation can work backwards, which hardly anyone accepts, the experience is illusory: the choice has already been made before the experience occurs, so the subjective phenomenon of choosing freely appears to be nothing more than a conscious rubber-stamping of something that was already happening without the agent consciously controlling it. That is no free choice.

Almost everything about Libet's experiment and the conclusions drawn from it and similar investigations is controversial both philosophically and scientifically, so if you want to know more, read on below. The point is that the experiments are interesting for two reasons. First, they simply demonstrate that the investigation of the phenomena of freedom is now in fair part an empirical matter: at any rate, empirical methods and experiments are now clearly germane to the freedom question. Secondly however, while in theory we could at some point see experimental results which more conclusively demonstrate the causal irrelevance of what appear to us subjectively as conscious choosing phenomena, the issue is still not finally settled, because the results are subject to more than one possible interpretation. Physicalists, who, like Frank Jackson in this volume, consider the totality of facts about the world to be exhausted by the totality of physical facts, would hail such a result as finally demonstrating that the experience of freedom is illusory. Libertarians on the other hand could take the result as demonstrating that physicalism cannot be the whole story: the brain events investigated by Libet cannot be all there is to the subjective phenomenon of choosing and initiating action, which libertarians typically take to be causes or partial causes of events both within and outside the chooser.

5. Moral consequences

It is widely held that if there is no freedom, then our understanding of human beings and our moral appraisal of human behaviour are due for the most radical upheaval ever. Essentially we should have to consider morality, which holds people personally and morally responsible for their free actions, and so to be subject to praise and blame, just reward and punishment for their actions, to have been founded on a huge mistake. The supporters of freedom consider this so outrageous and unlikely as to amount almost in itself to an argument for freedom. In addition to the almost universally upheld freedom assumption, practically our whole understanding of people, society and morality would also be largely illusory. It is hard, they say, to even think through coherently how we should (*should*?) restructure and revise our thinking. Would we lapse into barbaric anarchy and amorality? Would society, would humanity survive? Would the concept of freedom be somehow eradicated from our thinking, as some advocate it should be?

Compatibilists might take just this prospect as a spur to preserve the freedom assumption whatever the science tells us. But this tolerance is not infinitely stretchable: compatibilists are as bound by the scientific results as anyone else. Their only resorts if the chips were down for freedom would be to accede to libertarianism and accept that the scientific story is incomplete, or else to flip the other way and seek to build a morality compatible with hard determinism, instancing Spinoza and the Stoics as evidence that this is not itself a hopeless enterprise. Whether such attempts would or would not be coherent is precisely one of the points in dispute.

6. Towards a solution

The problem of free will is still unsolved, and consensus is far from being achieved. Everyone who thinks seriously about the subject has his or her own opinion. My own views for instance are broadly physicalist and compatibilist, though I am not a determinist, and I consider the roles of coincidence and chance in the texture of events to have been inadequately explored. However, as this volume attests, interest in the problem among philosophers, and the variety and sophistication of solutions put forward, have notably increased in recent years, and it is not least thanks to the stimulus of the scientific experiments by Libet and others. The study of what freedom is or may be, and whether it exists, is entering a new and exciting phase, where conceptual analysis can be assisted and complemented by empirical research. We are discovering new facts about the workings of the brain at an unprecedented rate, which hold out the promise that some or all of the mysteries surrounding the operation of free will can be resolved, and the controversy resolved. At the very least we can expect much food for thought and debate, and that is what this volume with its survey of the latest results and opinions aims to further.

Figures

1.1 Cortical structures involved in generation of voluntary action.   11
1.2 Transcranial magnetic stimulation over the motor cortex can
     be used to cause involuntary movements of the hand.   13
1.3 Schematic representation of the Libet experiment.   17
1.4 The design of Wenke and colleagues' experiment.   21

Contributors

Harald Atmanspacher is Head of the Department of Theory and Data Analysis, Institute for Frontier Areas of Psychology, Freiburg.

Tim Bayne is University Lecturer in the Philosophy of Mind at the University of Oxford and a Fellow of St Catherine's College.

R.A. Duff is Emeritus Professor of Philosophy, University of Stirling, Professor of Law, University of Minnesota, and a Fellow of the British Academy.

Solomon Feferman is Emeritus Professor of Mathematics and Philosophy, Stanford University.

Patrick Haggard is Professor of Cognitive Neuroscience, University College London.

Frank Jackson holds fractional research appointments at the Australian National University and La Trobe University, and is Visiting Professor of Philosophy, Princeton University, and a Corresponding Fellow of the British Academy.

J.R. Lucas was a tutorial Fellow of Merton College, Oxford, and is a Fellow of the British Academy.

Howard Robinson is University Professor, Department of Philosophy, Central European University, Budapest.

Stefan Rotter is Professor of Computational Neuroscience, Faculty of Biology, University of Freiberg, and Bernstein Centre, Freiberg.

Peter Simons is Professor of Philosophy, Trinity College, Dublin, and a Fellow of the British Academy.

Helen Steward is Senior Lecturer in Philosophy, University of Leeds.

Galen Strawson is Professor of Philosophy, University of Reading.

Richard Swinburne is Emeritus Nolloth Professor of the Philosophy of the Christian Religion, University of Oxford, and a Fellow of the British Academy.

Introduction

Plan of the volume

RICHARD SWINBURNE

The theme of this volume is the extent to which humans have a free choice of which actions to perform, and what kind of free choice would make them morally responsible for their actions. Earlier versions of six of the papers (those of Haggard, Bayne, Jackson, myself, Strawson, and Steward) were presented at a British Academy symposium in July 2010. The other five papers were written specially for this volume.[1] All the topics discussed in this volume have been the subject of very considerable recent philosophical and scientific discussions, and so – for those unfamiliar with these discussions, or with the scientific background – I have added to the volume a brief guide to background reading.

As Peter Simons's Foreword points out, it is only in the last few decades that neuroscientists have done experimental work, made possible by new investigative tools, which has begun to illuminate the processes which occur in our brains when we decide to perform some intentional action. The results of experiments of a kind pioneered by Benjamin Libet in the 1980s and 1990s have seemed to many neuroscientists to show that our decisions do not cause our actions, but that our actions are caused by brain events prior to our decisions to cause them; and that the same brain events also cause us to think (falsely) that our decisions really cause the actions. The results of the latest neuroscientific experiments are described in the

---

[1] I am most grateful to all who helped with the preparation of this volume, including all those who reviewed authors' first drafts and suggested improvements to them. I am especially grateful to Peter Simons, the current chair of the British Academy philosophy section which sponsored the symposium, for his help in the organization of the symposium, as well as in the preparation of the volume.

first paper in this volume by Patrick Haggard, who has led much of the work in this field, subsequent to Libet. The paper of Tim Bayne considers whether this work establishes nearly as much as its popularizers have claimed. Many philosophers hold that every conscious event is so closely connected to the brain event which gives rise to it, that perhaps a decision (a conscious event) can cause an action in virtue of the brain event which is its 'basis' causing the action. Bayne points out that, whether or not there is enough of a close tie to make this possible, the most that any experiments done so far have shown is that a brain event of a particular kind studied is a necessary condition for the occurrence of a decision, but not that it is sufficient by itself to cause the decision. (The brain event of this kind might be the basis of an urge or inclination to act, which might not always lead to a decision to act.) Finally he points out that while actions of the kind studied by neuroscientists are ones which – we believe – are caused by decisions, these decisions concern trivial matters (e.g. at exactly which moment in the course of a 20-second period to move a hand), and so are not decisions of the kind for which we hold people morally responsible.

How we are to assess this experimental work depends much on what it is for a brain event to be the 'basis' of a conscious event. On one view – physicalism – conscious events simply are brain events; and so all that neuroscience could show is which of our brain events are our decisions. (This view does seem to require a solution to what Bayne calls the 'causal exclusion problem'.) On a rival view – property (or event) dualism – there are events of two kinds: physical events (including brain events) and (separate from them) conscious events. A brain event is then only the 'basis' of a conscious event insofar as the former causes the latter. It is because many neuroscientists think of there being events of two separate kinds, that they think that their experimental results tend to show that conscious events do not cause brain events – a view called 'epiphenomenalism'. In his paper, Frank Jackson gives a brief argument for physicalism, and then draws out the consequences of physicalism for human free will. He argues that, given physicalism, human decisions are brain events caused (perhaps, if the fundamental laws of nature are indeterministic, to some small extent randomly) by previous brain events; and he then points out that this does not suggest a picture of humans as ultimately responsible for their actions. In my own paper I argue briefly for property dualism (arguments developed also by Howard Robinson in his paper), and I also give a brief argument against interpreting the experimental results as supporting epiphenomenalism. But if I am right about this, and there is good reason to believe that our decisions often cause brain events and thereby our intentional actions, there remains the crucial issue of whether our decisions are themselves caused (in the sense of 'fully predetermined') by brain events.

Property dualism together with the rejection of epiphenomenalism suggests the following picture of how human decisions are formed: brain events cause conscious events (in the form of beliefs, desires, sensations and thoughts), these interact and finally cause the formation of a decision; this then causes the brain event which causes the intentional action. Any laws at work in this process (of brain events interacting with conscious events) would be of a quite different kind from those studied by the physical sciences; and there would be an enormous number of separate laws, each of which applied to a different situation. In the case of someone reaching a decision about what to do in a situation where they have conflicting desires, value beliefs, and so on, the law governing what will happen will apply only to this situation (and could not have been previously tested). Consequently no one could ever have enough evidence to show that the formation of such a decision was in accord with a deterministic law rather than an indeterministic law. The fact that human decisions could not always be predicted with total accuracy is compatible either with those decisions being fully caused but not fully predictable, or with them not being fully caused. So, I argue in my paper, it may well be that brain events do not fully cause our decisions.

If our decisions cannot be predicted with perfect accuracy, and our decisions cause brain events, then brain events cannot be predicted with perfect accuracy; and if our decisions are not fully caused, then the brain events which they cause cannot be caused fully by other brain events. In that case brain processes cannot be fully deterministic (that is, the laws of nature do not always determine which brain event is followed by which other brain event). Many physicists believe that quantum theory has shown that the fundamental physical laws are not fully deterministic. Normally of course indeterminism on a small scale will average out so as to produce virtual determinism on the large scale. For example, if it were a totally indeterministic matter whether a coin landed heads or tails – if there was a physical probability (an inbuilt bias) of a half that the coin would land heads each time it was tossed, and a probability of a half that it would land tails – then in a million tosses, it would be very probable indeed that the proportion of tosses of heads would be very close to a half. But it is possible to have large-scale processes whose outcome is determined by very small-scale processes; for example scientists could construct a hydrogen bomb such that whether it exploded or not was determined by whether some atom which had a physical probability of one half of decaying within an hour, decayed within that time. Then it would be a totally chance matter (with a physical probability of a half) whether the bomb would explode or not. So the question arises of whether the brain is the kind of system in which small indeterminacies average out and make virtually no difference to which intentional actions we do, or whether it is a system in which small-scale events not fully determined by previous brain events (and so perhaps caused by uncaused decisions) cause our intentional actions. This issue is discussed in the paper by Harald Atmanspacher and Stefan Rotter. They argue that we do not at present know enough about how the brain works, to be able to answer this question.

Another discovery of modern science which has been thought to be crucially relevant to this issue is a theorem of mathematics, Gödel's incompleteness theorem. This states that for any consistent formal system (that is, a system with axioms and rules of inference which do not entail any contradiction) containing arithmetic, there will always be some well-formed formula which is neither provable nor disprovable from the axioms of the system. Now if all conscious events were caused by brain events in accord with deterministic laws, we could represent this process by a formal system in which the axioms describe the initial brain events and the rules of inference describe the causal laws governing the development of brain events and the production by them of conscious events. Among these conscious events would be mathematicians' awareness of mathematical truths. J.R. Lucas has argued over many years (with limited support from some others) that Gödel's theorem shows that there could not be such deterministic laws, since that theorem has the consequence that there will always be some mathematical formulas which a mathematician can see to be true (and would be provably true in some higher-level formal system), but which the system governing his brain would not cause him to see to be true. Hence, the argument goes, human mathematical calculations are not fully determined by brain events. In his paper, Solomon Feferman argues against drawing this conclusion from Gödel's theorem; and J.R. Lucas makes a brief reply to that paper.

Still, the mere fact that a decision has not been fully caused by previous events (whether conscious or physical) does seem to suggest that these are simply random events for which a person cannot be properly held morally responsible. That is the argument presented by Galen Strawson in his paper. He claims that, whatever the laws governing the formations of our decisions, it is simply not possible that a person can be morally responsible for their actions. For either they are caused to do what they do by events outside their control, or their actions are the result of random processes. In her paper, Helen Steward acknowledges that if people are morally responsible for their actions it must at least be the case that they themselves, not merely conscious of physical events which happen in them, cause their actions. They must be active agents. She argues that we are such agents, and agents of a particular kind such as to give us a choice of which actions to perform, independently of the prior causes which influence us. Even so, perhaps we can only be held responsible for our past actions if we are not merely agents but agents who remain essentially the same over time. But human bodies are continually changing, and perhaps the only way in which we can be unchanging agents over time is if we are essentially not bodies, but immaterial souls in control of bodies. This view that humans are essentially immaterial souls, and that conscious events are events in that soul, is the view known as substance dualism. In his paper, Howard Robinson considers the arguments for and against this view. He claims that such questions as 'Would I have existed if my mother's egg had been fertilized by a different though genetically identical sperm from my father?' must have a sharp yes-or-no answer (though we may not be able to discover what the answer is), but that they would not have a sharp answer if being me consisted simply in being made of similar genetic material and having a similar conscious life.

The criminal law punishes people for and only for breaches of the law for which they are morally responsible. However, if we come to believe that humans are not ultimately morally responsible for their actions, would not our understanding of the rationale for what we now call their 'punishment' have to change? Why should it remain relevant to this process whether the person to be 'punished' had or had not committed a crime at all? The only considerations relevant to punishing someone might seem to be whether 'punishment' would prevent them committing a crime while being punished, reform them so that they did not commit crimes in future, or deter others from committing similar crimes. R.A. Duff argues against that view. He claims that so long as a criminal is capable of understanding and responding to the reasons why the law defines some act as a crime, then if we punish him only if he has committed the crime that emphasizes that it matters that he committed the crime, and so we treat him as a member of the normative community.

5

The papers of Strawson, Steward, and Duff thus represent three different views on the compatibility of physical determinism and our responsibility for our actions. Steward is an 'incompatibilist', holding that physical determinism would rule out moral responsibility, Duff is a 'compatibilist' (with respect to the criminal law), claiming that we could still rationally hold people legally responsible for their actions even if physical determinism were true, while Strawson claims that we could never be (ultimately) responsible for our actions, whether or not determinism is true.

1
Does brain science change our view of free will?

PATRICK HAGGARD[1]

1. Introduction

Free will represents a frontier for Neuroscience. A neuroscientific account of free will would require a reductive explanation of something that we recognize as an important, even constitutive feature of ourselves as persons. In recent years, the possibility of a neuroscientific account of free will has attracted growing public attention. It now rivals classic neurophilosophical topics such as visual awareness in terms of public visibility. This may be partly because free will has a more direct importance to society than visual awareness: an idea of free will underlies the concept of responsibility for action, on which our legal and social systems depend. Our growing understanding of the brain mechanisms of human agency has brought two ways of thinking about persons into apparent conflict. The neuroscientific view focuses primarily on brain mechanisms: behaviour, decision and individual consciousness are all consequences of these mechanisms. On the other hand, the everyday, folk-psychological concept of free will is based on our first-person experience of choosing and acting, plus a strong component of social-cultural norms. These two views conflict in a number of ways. For example, if neuroscience were to invalidate our concept of free will, what would be the effects on our societies and ways of life? Less dramatically, if neuroscience were to show that free will resulted from a specific brain mechanism, how would society respond to actions of an individual with damage to that mechanism?

[1] Preparation of this chapter was supported by a Research Fellowship from the Leverhulme Trust, and by an ESF/ESRC research grant on 'Intentional inhibition'.

To begin, I want to declare two specific perspectives that this chapter will take. First, I will consider free will from the perspective of the human *motor* system. This view is justified because the problem of free will is a problem about the causes of our motor *actions*. For example, we might want to know what led to a person pulling a trigger, choosing to utter a particular word, or even eating a particular apple. The approach from motor neuroscience begins with the fact of the motor action, and then traces back to identify the particular stages of information processing in the brain that eventually produced it. The argument progresses backwards from actions towards intentions, the reverse of the natural process that leads forwards from intentions towards actions. The reason for this direction is simply methodological. Neuroscience needs to begin with the measurable facts of behaviours. Second, I will not engage with the philosophical debate over whether free will is compatible with determinism or not. Neuroscience is fundamentally deterministic, in its methods, its assumptions and its outlooks. Therefore, the perspective from neuroscience requires that free will must either be compatible with determinism, or we do not have free will.

This chapter explores the interaction between neuroscience and free will in the following way. First, I will consider how freely willed actions should be defined. Second I will outline our current understanding of brain mechanisms preceding action, showing in what respects these mechanisms meet the philosophical criteria for freely willed action, and in what respects they do not. Finally, I will end by concluding that the philosophical criteria themselves are based on two underlying psychological facts: human action involves complex mappings between environmental stimuli and goal-directed responses, and human action is associated with a range of quasi-perceptual experiences, classically called 'motor attention'. These facts lie at the heart of our concept of conscious free will, and are directly related to the recent evolutionary development of the brain's frontal lobes.

2. Criteria for freely willed action

These criteria for free will come from philosophy and folk psychology. We may therefore ask how these criteria fit with our understanding of the brain. If there are any freely willed actions, how are they generated in the brain?

Two characteristic features seem to define freely willed actions. The first is the well-known 'Could have done otherwise' feature. In philosophy, this is often used as a test of compatibility with determinism: given the pre-existing state of the universe, could the action have occurred differently, or not occurred at all? Movement neuroscience understands 'could have done otherwise' in terms of a process of action selection, that arbitrates between competing action alternatives, and selects the one that eventually occurs. So, for example, if we have evidence from brain measurements that two different action alternatives were developed in parallel, up to a late stage just before action execution, and if we can identify the particular brain process that led to one action alternative being selected and the other deselected, we may say that the person, or animal, could have done otherwise. The mechanism of selection itself may in fact be deterministic: that does not concern us here. The important point from a neuroscientific perspective is that the system supports a range of possible actions, rather than just one, and contains a mechanism that selects definite actions, and eliminates alternative possibilities. Whether this is also a reasonable philosophical interpretation of 'could have done otherwise' is less clear. Much depends on the meaning of the word 'could'. The philosophical interpretation often carries the idea of whether an alternative possibility was open or not, given laws of nature and initial conditions. The neuro-scientific interpretation emphasizes only that alternative possibilities were *represented* in the mind/brain, and does not comment on whether those possibilities could have actually happened.

The second characteristic of freely willed actions is more elusive. Our concept of free will implies a conscious subject, who initiates the action, and is therefore responsible for it. That is, there must be an 'I' who could have done otherwise. Moreover, the agent must be conscious that they are making the action. We will therefore refer to this as the *subjective agency*. Therefore, concepts of free will do not apply when this subjectivity is absent, for example to a person who acts in their sleep, or to an involuntary movement, such as a simple reflex. Neuroscience treats the 'I' as being synonymous with an individual's brain. Therefore, it approaches the awareness of action by distinguishing between those action-related processes in the brain that typically generate conscious experiences, and those that do not. The neuroscientific view of *conscious* free will accordingly pays particular attention to the relation between brain activity in the motor system and the experience of intending and performing an action.

3. Methodology: how can volition be studied experimentally?

Before examining the contributions of brain science to our understanding of volition, we must first ask about methodology: how can volition be studied experimentally?

Voluntary action often appears to vanish in the controlled conditions of the experimental laboratory. In particular, good experiments require good specification of the stimulus, and clear instructions, which both seem incompatible with free will. Thus, an experiment with the instruction 'have free will, now!' seems absurd. The concept of free will implies that the participant themselves, rather than the experimental situation, brings the triggers for action. This impasse persuades many that volition cannot be studied scientifically. Indeed the concept largely disappeared from psychology during the behaviourist and cognitivist periods (Skinner 1953).

Experimentalists have generally escaped from this impasse by comparing 'fixed' actions, where an external stimulus specifies which of a range of possible actions a person must make, with 'free' actions, where the participant must generate this information for themselves. For example, Deiber *et al.* (1991) compared a condition in which participants were instructed to move a joystick repetitively in a single direction, with a condition in which the participants themselves selected the direction. These conditions both involved a physical movement, but only the second condition involved free selection from among action alternatives. Thus, comparing the two conditions should reveal the brain activity associated with voluntary action selection.

This 'fixed vs. free action' approach provides an operational definition of volition, based on externally specified vs. internally generated information for action. It has come to dominate most neuroscientific studies. Brass and Haggard (2008) argued that decisions about whether to perform an action, what action to perform, and when to perform it, could all be understood within this framework. This operational definition does appear to satisfy the 'could have done otherwise' criterion for free will, as described above. However, the subjective agency component is still missing. Contrasting forced vs. free actions tells us nothing about the subjective aspect of action selection, nor about why and how one particular action was chosen over another. The participant clearly generates action information internally, but we do not know how. One speculation is that people try to choose randomly, in as unfixed a way as possible.

4. Brain pathways underlying voluntary action

The idea of voluntary action is intimately tied to the control of muscular movement. Wittgenstein asks: 'what is left over if I subtract the fact that my arm goes up from the fact that I raise my arm' (Wittgenstein 1953). Therefore, the logical way to explore the brain pathways underlying voluntary action is in the backwards direction, beginning with the brain area that sends the final motor command to the muscles: the primary motor cortex. The primary motor cortex forms a strip of tissue immediately in front of the central sulcus in each cerebral hemisphere (Figure 1.1).

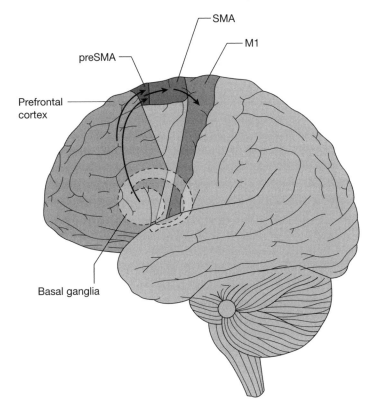

Figure 1.1 Cortical structures involved in generation of voluntary action.

M1 – primary motor cortex; SMA – supplementary motor area; preSMA – presupplementary motor area.

Source: P. Haggard (2008) 'Human volition: towards a neuroscience of will', *Nature Reviews Neuroscience*, 9: 934–46.

11

5. Primary motor cortex: the final common path

The primary motor cortex represents the major motor output area of the cerebrum: the neurons here communicate directly with the lower motorneurons in the spinal cord, and so are only one synapse away from the muscle itself. Interestingly, in recent years it has become possible to activate this area directly from outside the brain using transcranial magnetic stimulation (TMS). The brief magnetic field induced by the machine (Figure 1.2) passes through the skull, and causes neurons in the primary motor cortex to fire. About 20ms later, depending on the distance that the nervous impulse has to travel between brain and muscle, there is a small involuntary twitch of the hand, which can easily be measured using electrodes placed over relevant hand muscles. Interestingly, the experience of TMS is absolutely different from the experience of voluntary action: the participant quite distinctly feels that they are *being moved*, like a marionette is moved by pulling a string. This result immediately suggests that the primary motor cortex cannot form part of the brain mechanism underlying free will, because it does not satisfy our criterion of subjective agency. Indeed, if stronger TMS pulses are used to delay a voluntary action, the participant does not perceive that their action has been delayed. This suggests that the experience of voluntary action has been formed prior at an earlier stage (Haggard and Magno 1999).

The brain mechanisms of free will must therefore lie upstream, in the areas that initiate action, and not in the final output stage that dispatches motor commands to the muscles. Sherrington developed the classic physiological concept of a 'final common path' to action (Sherrington 1906). This states that, while there are many reasons and origins for our actions, they must ultimately converge on to the specific centres that cause our muscles to contract. The primary motor cortex is part of the final common path that leads to muscle. But the discussion above shows that our criteria of 'could have done otherwise' and subjective agency apply at the beginning of the final common path, closer to the reasons for action, and not at the end.

6. The supplementary motor areas underlying action selection

Therefore, pursuing the motor chain backwards from the primary motor cortex should help to identify brain areas underlying voluntary action.

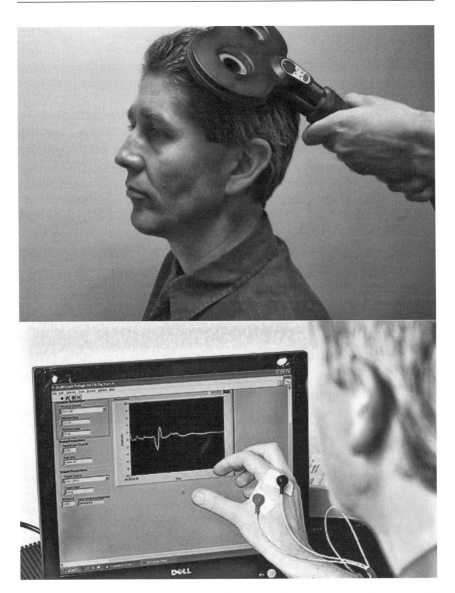

Figure 1.2 Transcranial magnetic stimulation over the motor cortex (upper panel) can be used to cause involuntary movements of the hand (lower panel). Note the electrodes placed over the muscles that primarily move the index finger. The deflection of the trace on the computer screen shows the twitch caused by the stimulation.

13

Here, there has been strong interest in areas immediately in front of the motor cortex, notably the supplementary motor area (often termed SMAproper) and the presupplementary motor area (preSMA). These areas will therefore be discussed in detail, because they seem to come close to our criterion.

To understand what a brain area does, neuroscientists often begin by noting how behaviour changes when the brain area is damaged. They then conclude that the function of the brain area corresponds to whatever the person or animal cannot do following localized damage. There is convincing evidence that the preSMA is involved in the selection of alternative actions, and that it achieves this selection through competition between action possibilities. A dramatic instance of this comes from Anarchic Hand Syndrome (AHS), which is occasionally seen after damage to the SMA in one hemisphere, particularly when the callosal fibres that carry signals between the two hemispheres are also damaged (Della Sala *et al.* 1991). In AHS, the patient performs unintended, goal-directed actions with the affected hand, as if the hand is driven by the immediate stimulus environment, rather than by the patient's personal will. Della Sala *et al.* give the compelling example of a patient saying that they will wait before taking a drink, because the beverage is probably still too hot to drink, while the affected hand compulsively moves towards the cup, to grasp it and bring it to the mouth. Patients with small lesions in this area, but without full AHS syndrome, are unusually responsive to subliminal priming by visual stimuli, as if their action is captured by current sensory information (Sumner *et al.* 2007).

The broad picture that emerges from this literature is of environmental stimuli that continually elicit competing responses. A major part of volition involves the inhibitory process of deselecting and suppressing those stimulus-triggered responses when they are inappropriate, or not useful for our current goals. Volition involves exerting internal motivations against external triggers for action.

7. Readiness potentials and action initiation

This data can explain the negative side of volition (why we do not constantly react), but says less about the positive side (why we sometimes perform actions that relate to our goals and wishes). Prior to voluntary actions, a characteristic pattern of electrical activity occurs in the SMA and preSMA. This can be recorded by electrodes on the frontal scalp as a 'readiness-potential', or gradual ramp-like increasing negative voltage, beginning 1s or more before action, and ending abruptly around the time of action itself (Kornhuber and Deecke 1965). Associated with growing premotor activity in frontal motor regions is a decrease in the spontaneous oscillatory activity of the same regions (Neuper *et al.* 2006). The conventional recording method, with electrodes placed on the scalp, does not have enough spatial resolution to identify precisely the underlying brain regions. However, intracortical recordings in neurosurgical patients with electrodes implanted in the brain confirm that the preSMA and SMAproper are involved (Ikeda *et al.* 1999).

In fact, recent neuroscientific data suggest that the positive process of initiating action is intimately linked with the process of action selection. Our 'could have done otherwise' criterion focuses on the situation where the person decides for themselves on the details of action. Interestingly, when a person is asked to freely choose between an action of the left or right hand, the later stages, but not the earlier stages, of the readiness potential show a greater voltage over the right and left hemispheres of the brain, respectively, reflecting the fact that muscular control of each side of the body depends on the motor areas in the contralateral hemisphere. Thus, the readiness potential is divided into an early non-lateralized phase, where the participant might presumably make either a left or right hand action, and a later, lateralized phase where these details of action are clearly more fixed.

Further evidence of a link to action selection comes from a recent study of premovement reduction in neural oscillations that accompanies the readiness potential (Tzagarakis *et al.* 2010). When participants were warned in advance that they would shortly have to move in one of several possible directions, the oscillatory power decreased as the number of possible movements decreased. For example, when participants knew they need select from only two possible movement directions, oscillatory power was more strongly reduced than when they were instructed that they would select from three.

The readiness potential, and associated reductions in neural oscillations, are important measures of the processes that *initiate* voluntary action. These processes can be divided into an early phase, which does not fully specify the details of which action will be performed, and a later phase, which contains more motor detail. The neural events that precede action can be seen either as an energization process that triggers action, or as a progressive reduction of uncertainty that selects a specific intended action, and deselects alternatives. While we experience our voluntary action as a positive process aiming towards a goal, neuroscientific data show that initiation of voluntary action is associated with a neural process of action selection. Indeed, to make a voluntary action may simply be to reduce the size of what has been called the 'response space', to converge on a single action. In terms of the 'could have done otherwise' criterion, there is strong evidence for a dynamic process that gradually evolves from a situation where multiple actions are possible, to a point of no return, where a single action enters the final common path in the primary motor cortex. One might say that in voluntary action, we move from a phase where we clearly could do otherwise, to a point where we are prisoners of our current motor command. We may be free before we act, but once we act, counterfactuals become irrelevant. Responsibility for action requires that actions, in the end, are objective, physical facts about the world.

8. Brain processes underlying the experience of volition

We now turn to the second criterion, of subjective agency. The phenomenology of volition is hard to study experimentally. We know when we make voluntary actions, and we certainly have no difficulty discriminating our voluntary actions from physically identical passive movements, as in Wittgenstein's example of raising one's hand. However, the feeling of voluntary action is often thin and elusive. I want to walk to the station, and I decide to do so, but I do not have an intense experience of volition with each step that takes me there.

The best-known approach to the conscious experience of willing is the chronometric approach, in which people are asked to report the moment in time when they experience the intention to initiate an action. Clearly, this method is restricted to the immediate, decisional aspect of volition that precedes action, and cannot capture general, longer-range aspects of volition, such as my intentions to be more generous to my friends, or to repaint the house. The classic experiment in this area is that of Benjamin Libet (Libet *et al.* 1983). Libet and colleagues asked participants to make simple voluntary actions at a time of their own choosing while watching a rotating clock hand. Participants were further asked to note the position of the clock hand at which they first felt the 'urge' to move. After the action itself, the clock stopped after a random delay, and the participant reported the previously-noted clock time. Readiness potentials were measured from the scalp throughout.

The result of the experiment is straightforward (Figure 1.3). The readiness potential began up to 1s or more before movement onset. The experience of willing the action occurred on average 206ms before movement onset. Libet argued that conscious will could not therefore be the cause of our actions, since the preparatory brain activity that produces the action begins before the conscious state of willing, and causes cannot precede their effects. Incidentally, this method of measuring perceived timing has attracted considerable criticism (see for example the commentaries to Libet's 1985 target article in *Behavioural and Brain Sciences*). However, these arguments have been extensively reviewed elsewhere. Here we focus principally on the relation between brain activity and subjective agency, rather than on the specific temporal order in the original experiment. From a neuroscientific perspective, Libet's result is actually

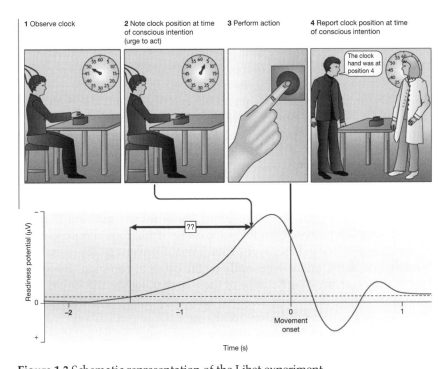

Figure 1.3 Schematic representation of the Libet experiment.

Source: P. Haggard (2005) 'Conscious intention and motor cognition', *Trends in Cognitive Sciences*, 9(6): 290–5.

unremarkable. Conscious experience is a product or aspect of some brain processes, and not an event independent of the brain. Although consciousness may be part of brain activity, consciousness cannot cause brain activity, nor can it cause actions.

An alternative reading of Libet's results might be that the readiness potential, once it reaches a particular threshold value, produces an experience of conscious intention-in-action, i.e., of being about to act. Haggard and Eimer (1999) used Mill's method of concomitant variation to investigate this possibility. They reasoned that, if the readiness potential is indeed the cause of the conscious experience of intention, actions where the participant happens, for whatever reason, to experience a very early intention, long before they act, should have earlier readiness potentials than trials where the participant happens to experience intention rather later, just prior to the action itself. This argument has an important advantage over temporal precedence arguments, namely that it is unaffected by systematic errors in subjective time perception.

In fact, Haggard and Eimer found no evidence for such a relation between readiness potentials and the perceived time of intention. Further, their experiment gave participants a free choice between making an action with the left or right hand. In contrast, Libet's original experiment involved only unimanual actions. As described above, when people choose between left- and right-hand actions, the readiness potential begins over both brain hemispheres, but then lateralizes to the hemisphere contralateral to the hand that will move. Haggard and Eimer found a significant relation between the onset time of the *lateralized* readiness potential and the time of conscious intention. When conscious intention occurred early, there was an early lateralization of the readiness potential. When conscious intention occurred late, just prior to the action itself, there was a late lateralization of the readiness potential. Taken together, these results indicate that the general readiness potential cannot be the cause of the conscious experience of intention reported in Libet-type experiments. Conversely, the lateralized readiness potential might potentially be the cause of conscious intention. This finding is interesting because of the role that it gives the 'could have done otherwise' criterion in subjective agency. Specifically, the experience of conscious intention-in-action occurs once the action has taken a concrete motor form, and therefore after the processes of action selection. The brain processes that choose from among a set of alternative actions seem to indicate the point at which conscious experience enters the chain of voluntary action.

18

9. Are experiences of intention just retrospective confabulations?

A common objection to these experiments is based on the comment that participants report the time intention only *after* they have acted. Therefore, they may reconstruct an experience of 'intention' as part of a narrative to explain the occurrence of the action. Intentions are really narrative reconstructions, or even confabulations, and play no part in the brain processes that lead to action. This view has been called 'The illusion of conscious will' (Wegner 2003). While the mind clearly does reconstruct such narratives, I want to highlight two pieces of evidence from neurophysiology and psychology, that the subjective experience of voluntary action cannot *only* be based on retrospective reconstruction.

The neurophysiological evidence comes from direct stimulation of the human brain. In some cases of drug-resistant epilepsy, electrode grids are placed directly on to the cortical surface for exploratory stimulation prior to neurosurgical treatment. In a typical exploratory session, the patient remains fully conscious while the neurosurgeon stimulates each electrode in the grid. In some patients, stimulation of the supplementary motor areas produces a feeling of an urge to move. The urge refers to a specific body part, such as the left hand (Fried *et al.* 1991). Crucially, the patient reports this feeling even when they do not know where or if they are being stimulated, and while remaining quite still. Since the patient has not actually moved, it is unclear how their experience could be based on a reconstructive narrative. Rather, it appears that the neurosurgeon has given them, quite involuntarily, an experience which seems like a percept of volition. The will is clearly not free in this case, since the neurosurgeon controls it. However, this data provides powerful evidence that the normal experience of subjective agency in volition is a direct consequence of activity in the supplementary motor areas.

In fact, neurosurgical findings of an urge to move are rather widespread in the brain. They have also been reported in the parietal lobes (Desmurget *et al.* 2009), with which the frontal motor areas are strongly interconnected (see below). For our purposes, the precise location does not matter. The important point is that an external intervention within a specific brain mechanism can generate an experience which has at least some of the key features of volition.

The second class of evidence comes from psychological studies investigating the perceived feeling of control. Wenke and colleagues (including the author) investigated whether the feeling of control over the effects of an action depends on the processes that initially select which action to make (Wenke *et al.* 2010). Since these action-selection processes necessarily occur before action itself, any such influence cannot be retrospective. Participants pressed a button with their left or right hand. In some trials, a large, clearly visible arrow showed which button to press. In other trials, a double-headed arrow appeared, meaning that they should freely choose for themselves which action to make. Without the participants' knowledge, the large arrows were preceded by invisibly brief presentations of smaller left- or right-pointing arrows. When participants were asked to freely choose between left and right keypresses, these 'subliminal priming' arrows were able to bias their 'free' choices, so that they were slightly more likely to choose the action compatible with the prime (e.g. left-priming arrow followed by a free choice to press the left key), rather than the incompatible action. When participants were explicitly instructed which key to press by the larger arrow, their reactions were slower and they were more likely to press the wrong key, if the large arrow had been preceded by an incompatible prime than by a compatible one. This shows that the priming arrows, though invisible, influenced the selection of the keypress actions, even when these were supposedly 'free'.

Each keypress was then followed by one of several possible colours appearing briefly on the screen. After many such trials, participants rank ordered all the different colours they had seen, in order of the level of control they had felt over the appearance of each colour. Crucially, we organized the colour patches so that subjects saw one set of colours when their action was compatible with the preceding subliminal prime, and another set of colours when their action was incompatible with the preceding prime.

The results showed that participants experienced a greater level of control over the colour when their action was compatible with a preceding subliminal prime than when it was incompatible. It is important to point out that the colour does not depend directly on the subliminal prime, but on whether the prime influences the subject's action selection processes or not. These results show that the initial, premovement process of selecting which action to perform makes an important contribution to the feeling of control over actions and their effects. Extrapolating to situations outside the laboratory, we can say that the sense of agency is not merely a post-hoc inference or confabulation to account for the actions that we find we have made. Rather, the sense of agency depends on the brain processes that lead

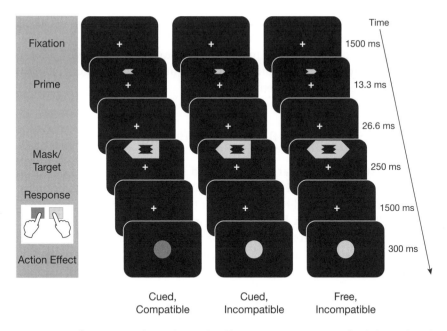

Figure 1.4 The design of Wenke and colleagues' experiment. The leftward and rightward primes presented at the start of each trial are not consciously seen by the participant, though they bias how they respond to the subsequent instruction given by the large arrow targets. More interestingly, participants feel a stronger sense of controlling the subsequent colour patch caused by their action when they act in a way that is compatible with the unseen primes.

Source: D. Wenke, S.M. Fleming and P. Haggard (2010) 'Subliminal priming of actions influences sense of control over effects of action', *Cognition*, 115(1): 26–38.

us to make this particular action, rather than any other action, in the first place.

Wenke *et al.*'s experiment did not specifically ask about the experience of intention or volition, but only about the experience of control over effects of action. Interestingly, however, the results show that a premovement process of voluntary choice is highly relevant to the experience of control. This experiment also has interesting societal implications. The sense of control is strongly linked to the social concept of responsibility. For example, the law assumes that an individual has a reasonable awareness of the consequences of their action at the time that they perform the action. Wenke *et al.*'s experiment suggests that the sense of control depends, at least partly, on the way that the plan for action is selected and formulated

21

in the mind, prior to the actual physical execution of action. In that sense, to accept responsibility for one's actions is simply for there to be a particular functional relation linking the mental and neural states that occur prior to action, the physical movements of the body, and the subsequent external events caused by those movements.

10. Free will as action flexibility and complexity

At the start of this chapter, we suggested that brain processes of action selection may correspond to the 'could have done otherwise' criterion for free will. This view leads to the thought that the characteristic feature of freely willed actions may be appropriate engagement of the brain's action-selection system. Neuroscience makes a very general distinction between simple reflex actions, in which a single stimulus elicits a stereotypical response, and the more complex case, where a range of stimuli are mapped on to a range of responses, with a contextual rule defining the mapping between the two. Everyday life offers ready examples of such mappings: a red light means 'don't walk' if I am trying to cross the road, but a rather similar red light appears above the lift to tell me that the lift is here, and I can walk into it. This kind of rule-based mapping offers a kind of flexibility of action that simply does not exist for simple orienting responses, such as looking towards the source of an unexpected loud noise. Moreover, an action selection mechanism is required to map stimulus and context to action in the correct way. In humans, the rules mapping stimuli and contexts to appropriate responses can become extremely complex. The number of stimuli and contexts that are mapped is also extremely large. The key brain regions for implementing complex, flexible mappings are in the frontal and prefrontal cortex. The dorsolateral prefrontal cortex houses a form of working memory, that links the set of available stimuli to alternative actions, and possibly also selects from them. The medial frontal and prefrontal cortices seem to house a competitive selection from among a wider set of possible actions, stored in long-term memory.

Perhaps the key feature of 'free will' is simply the complexity and flexibility of the mappings in the brain that produce some of our actions. For freely willed actions, there often seems to be no obvious external cause or stimulus to explain why the agent performed one action rather than another. Dualist philosophies classically revert to some form of agent causation at this point. Nothing in the external world caused the action.

Rather, the agent themselves caused the action, even though they could have done otherwise. The neuroscientific view, in contrast, would suggest that everything, rather than nothing, is the cause of our freely willed actions. Such actions may simply be responses to a situation whose complexity seems to exhaust our attempt to identify a simple cause for each action.

11. Conclusion

Brain science will change our notion of free will in several ways. Most importantly, it will show that free will is not a special, transcendental faculty. Rather, it is a term given to the operations of a set of brain processes in the frontal lobes of humans, and possibly some other primates. More specifically, brain science will progress the philosophical analysis of free will in three distinct ways. First, it will show that the 'could have done otherwise' criterion of free will refers to the engagement of a neural memory buffer that stores possible alternative actions. Second, it will show that the conscious experience of initiating and controlling is a quasi-perceptual process resulting from neural activity in specific frontal and parietal brain areas. Finally, it will provide an alternative way of thinking about agent-causation. Conventional metaphysics of free will invokes an 'I' to consciously initiate willed actions. Instead, neuroscience emphasizes that 'freely willed' actions may simply have a more complex set of causes than simpler actions. Nevertheless, these causes are not special in any particular sense: they simply reflect the flexibility and complexity of our response to our environment.

References

Brass, M. and Haggard, P. (2008) The what, when, whether model of intentional action, *The Neuroscientist*, 14(4): 319–25.

Deiber, M.P., Passingham, R.E., Colebatch, J.G., Friston, K.J., Nixon, P.D. and Frackowiak, R.S. (1991) Cortical areas and the selection of movement: a study with positron emission tomography, *Experimental Brain Research. Experimentelle Hirnforschung. Expérimentation Cérébrale*, 84(2): 393–402.

Della Sala, S., Marchetti, C. and Spinnler, H. (1991) Right-sided anarchic (alien) hand: a longitudinal study, *Neuropsychologia*, 29(11): 1113–27.

Desmurget, M., Reilly, K.T., Richard, N., Szathmari, A., Mottolese, C. and Sirigu, A. (2009) Movement intention after parietal cortex stimulation in humans, *Science* (New York), 324(5928): 811–13.

Fried, I., Katz, A., McCarthy, G., Sass, K.J., Williamson, P., Spencer, S.S. and Spencer, D.D. (1991) Functional organization of human supplementary motor cortex studied by electrical stimulation, *The Journal of Neuroscience: The Official Journal of the Society for Neuroscience*, 11(11): 3656–66.

Haggard, P. and Eimer, M. (1999) On the relation between brain potentials and the awareness of voluntary movements, *Experimental Brain Research. Experimentelle Hirnforschung. Expérimentation Cérébrale*, 126(1): 128–33.

Haggard, P. and Magno, E. (1999) Localising awareness of action with transcranial magnetic stimulation, *Experimental Brain Research. Experimentelle Hirnforschung. Expérimentation Cérébrale*, 127(1): 102–7.

Haggard, P. (2005) Conscious intention and motor cognition, *Trends in Cognitive Sciences*, 9(6): 290–5.

Ikeda, A., Yazawa, S., Kunieda, T., Ohara, S., Terada, K., Mikuni, N. and Nagamine, T. (1999) Cognitive motor control in human pre-supplementary motor area studied by subdural recording of discrimination / selection-related potentials, *Brain: A Journal of Neurology*, 122(5): 915–31.

Kornhuber, H.H. and Deecke, L. (1965) [Changes in the brain potential in voluntary movements and passive movements in man: readiness potential and reafferent potentials], *Pflügers Archiv Für Die Gesamte Physiologie Des Menschen Und Der Tiere*, 284: 1–17.

Libet, B., Gleason, C.A., Wright, E.W. and Pearl, D.K. (1983) Time of conscious intention to act in relation to onset of cerebral activity (readiness-potential): the unconscious initiation of a freely voluntary act, *Brain: A Journal of Neurology*, 106(3): 623–42.

Neuper, C., Wörtz, M. and Pfurtscheller, G. (2006) ERD / ERS patterns reflecting sensorimotor activation and deactivation, *Progress in Brain Research*, 159: 211–22.

Sherrington, C.S. (1906) *The Integrative Action of the Nervous System*, New York: Charles Scribner's Sons.

Skinner, B.F. (1953) *Science and Human Behavior*, New York: Macmillan.

Sumner, P., Nachev, P., Morris, P., Peters, A.M., Jackson, S.R., Kennard, C. and Husain, M. (2007) Human medial frontal cortex mediates unconscious inhibition of voluntary action, *Neuron*, 54(5): 697–711.

Tzagarakis, C., Ince, N.F., Leuthold, A.C. and Pellizzer, G. (2010) Beta-band activity during motor planning reflects response uncertainty, *The Journal of Neuroscience: The Official Journal of the Society for Neuroscience*, 30(34): 11270–7.

Wegner, D.M. (2003) *The Illusion of Conscious Will* (First Edition), Cambridge, MA: MIT Press.

Wenke, D., Fleming, S.M. and Haggard, P. (2010) Subliminal priming of actions influences sense of control over effects of action, *Cognition*, 115(1): 26–38.

Wittgenstein, L. (1953) *Philosophical Investigations*, Oxford: Blackwell.

2
Libet and the case
for free will scepticism

TIM BAYNE[1]

1. Introduction

Free will sceptics claim that we do not possess free will – or at least, that we do not possess nearly as much free will as we think we do. Some free will sceptics hold that the very notion of free will is incoherent, and that no being could possibly possess free will (Strawson, this volume). Others allow that the notion of free will is coherent, but hold that features of our cognitive architecture prevent us from possessing free will. My concern in this chapter is with views of the second kind. According to an increasingly influential line of thought, our common-sense commitment to the existence of free will is threatened in unique ways by what we are learning from the sciences of human agency.

We can group such threats into two categories. One kind of threat purports to 'undercut' or 'undermine' our reasons for belief in free will. To develop a successful objection of this kind one must first identify the basis on which we believe in free will, and then show that this basis is unlikely to yield true beliefs. Although they are not without interest, undercutting objections have not been at the heart of the contemporary case for free will scepticism. Instead, those who invoke the sciences of agency to motivate free will scepticism typically mount rebutting objections to free will. Whereas undercutting arguments attempt to undermine our evidence for free will, rebutting arguments provide what are alleged to be positive reasons against free will.

[1] I am very grateful to Neil Levy, Richard Swinburne and an anonymous reviewer for their helpful comments on earlier versions of this paper.

This chapter examines what is arguably the most influential rebutting objection in the current literature, an objection that appeals to Benjamin Libet's studies concerning the neural basis of agency.[2] Although Libet himself stopped short of endorsing free will scepticism on the basis of his results, other theorists have not been so cautious, and his work is often said to show that we lack free will.[3] I will argue that Libet's findings show no such thing. However, Libet's experiments do raise a number of interesting and important questions for accounts of free will. In particular, Libet's experiments raise challenging questions about the analysis of the concept of free will. In order to determine whether brain science supports free will scepticism we need not only to understand the relevant brain science, we also need to understand just what the common-sense or folk notion of free will commits us to. As we will see, the latter requirement may be as difficult to meet as the former one is.

2. The Libet paradigm

Let us begin with an overview of Libet's experimental paradigm (Libet 1985; Libet *et al.* 1983). Subjects are told to perform some simple motor action, such as flexing their wrist, at the moment of their choosing within a specified period of time (say, 30 seconds), and that this action should be performed 'spontaneously'. At the same time, they are instructed to monitor their agentive experiences, and to identify the time at which they were first aware of the 'decision', 'urge' or 'intention' – Libet used these terms interchangeably – to act. Subjects do this by watching a clock face with a dial that rotates rapidly (once every 2560ms). Libet referred to the judgement of the time of their 'decision' ('urge', 'intention') to act as the 'W judgement'. While subjects were both acting and monitoring their decisions (urges, intentions) to act, Libet used an EEG to measure their neural activity. These measurements revealed preparatory brain activity – what Libet called a type II readiness potential (RP) – prior to the action.

---

[2] The literature on Libet's experiments is large and expanding. Useful entry points into it are provided by Banks and Pockett (2007), Gomes (1999), Haggard (2008), Levy (2005), Mele (2009) and the chapters in Sinnott-Armstrong and Nadel (2011).

[3] For examples of free will scepticism that appeal to Libet's work, see Banks and Isham (2011), Hallett (2007), Pockett (2004), Roediger *et al.* (2008), Spence (2009) and Wegner (2002).

The critical question in which Libet was interested concerned the temporal relationship between RP and the content of the subjects' W judgements.[4]

The EEG revealed that the RP preceded the subjects' actions by about 550ms. However, on average subjects reported that they felt that they had decided to move only 200ms prior to the action (dating that point to the onset of muscle activity initiating the movement). In other words, there appeared to be a gap of about 350ms between the RP and the point at which subjects claimed to be aware of their decision (urge, intention) to act. In fact, Libet claimed that this gap was around 400ms in length, for he argued that subjects were aware of their agentive decisions only 150ms (rather than 200ms) before they acted. His argument for this claim appealed to the fact that subjects appear to misjudge the point at which a tactile stimulation is applied to the body by about 50ms. Although I have doubts about whether this correction is justified, not a great deal turns on this issue in the present context and I will accept it here.

As a number of commentators have pointed out, Libet's paradigm is subject to a number of methodological problems (see, e.g., the commentaries on Libet 1985). To take just one example of these problems, Libet's paradigm requires subjects to divide their attention between the position of the clock face and their own agency. The demand to divide one's attention between two perceptual streams in this way is a notorious source of error in temporal order judgements. Despite these difficulties, Libet's basic findings have been replicated by a number of laboratories using studies that are free of these methodological difficulties.[5] Although there is some variability between studies, the claim that 'Libet-actions' – that is, simple and (relatively) spontaneous motor actions – involve an RP whose onset precedes the time of the subjects' W judgement by about 400ms or so is largely undisputed. What is in dispute is the implications of these results for questions concerning free will.

Although Libet took his experiments to put pressure on the folk notion of free will he did not think that they established free will scepticism, for he argued that the gap of 150ms between the agent's conscious decision

---

[4] Libet distinguishes two types of readiness potentials. Actions that are performed spontaneously (as the actions studied in this experiment were said to be) involve a type II RP, whereas pre-planned actions exhibit what Libet calls a type I RP. Type I RPs can be seen up to 1500ms prior to the action (Libet *et al.* 1982; Trevena and Miller 2002).

[5] For replication of Libet's basic findings, see Haggard and Eimer (1999); Keller and Heckhausen (1990); Lau *et al.* (2004) and Trevena and Miller (2002).

and the onset of the action allowed for a kind of free will in the form of conscious veto. However, many theorists have seen in Libet's work the death-knell of free will. In their review of his work, Banks and Pockett (2007: 658) describe Libet's experiments as providing 'the first direct neuro-physiological evidence in support of [the idea that perceived freedom of action is an illusion]'.

Unfortunately, few sceptics have said exactly how Libet's data is supposed to undermine free will. The central sceptical worry clearly involves the thought that the neural data reveals conscious decisions to be epiphenomenal, but there is more than one way in which this general concern can be corralled into an argument against free will. The following argument seems to me to be closest to capturing the heart of the sceptical appeal to Libet's results, and I will structure my discussion around it.

(1) The actions studied in the Libet paradigm are not initiated by conscious decisions but are instead initiated by the RP.
(2) In order to exemplify free will an action must be initiated by a conscious decision.
(3) So, the actions studied in the Libet paradigm are not freely willed. [From (1) and (2).]
(4) Actions studied in the Libet paradigm are central exemplars of free will (as intuitively understood), and so if these actions are not freely willed then no (or at least very few) actions are freely willed.
(5) So no (or very few) human actions are freely willed. [From (3) and (4).]

I will refer to this as 'the sceptical argument'. The sceptical argument is valid, so if it is to be resisted we need to reject one (or more) of its premises. I will examine the premises in reverse order, beginning with (4).

3. The scope of free will

Are the actions that form the focus of the sceptical argument – 'Libet-actions' – paradigm examples of our intuitive notion of free will? Libet himself had no doubts about the answer to this question, for he took himself to have studied an 'incontrovertible and ideal example of a fully endogenous and "freely voluntary" act' (Libet *et al.* 1983: 640). However, not everyone shares this view. Adina Roskies, for example, claims that Libet-actions are at best 'degenerate' examples of free will, and suggests that we ought to focus on actions that are grounded in our reasons and motivations if we are interested in 'how awareness and action are related insofar as they bear on freedom and responsibility' (Roskies 2011: 19).

To get to the bottom of this issue we need to characterize the kind of actions that are performed in Libet experiments, and to do that we need a taxonomy for actions. Most fundamentally, we can distinguish *automatic* actions from *willed* actions. This distinction can be roughly mapped on to the distinction between endogenous and exogenous actions (Haggard, this volume), and should be thought of as a distinction between two ends of a continuum rather than a distinction between two discrete categories. Automatic actions flow directly from the agent's standing intentions and pre-potent action routines. Many of our everyday actions – washing the dishes, answering the telephone, reaching for a door handle – are automatic. Our awareness of various features of our environment together with over-learned action schemas conspire to trigger the appropriate intentions with only the minimal participation of conscious deliberation or decision on the part of the agent.

Willed actions, by contrast, require the intervention of executive processes – they require acts of choice and decision. We can distinguish between different forms of willed agency. Consider the experience of finding oneself in a restaurant confronted by a number of equally attractive – or, as the case may be, unattractive – options on the menu. One needs to make a choice, but it does not matter what one orders. Without an act of will one would, like Buridan's ass, starve to death, but little rides on the content of what one wills. We might call this *disinterested agency*. Exercises in disinterested agency must be distinguished from exercises in *deliberative agency*, where the latter involves the interrogation of one's reasons and commitments. Consider Sartre's case of the young man who must choose whether to look after his aged mother or join the resistance. Here, the function of decision-making is not to select from amongst a range of options between which one is relatively indifferent, but to draw on one's reasons in making a good decision. Should one join the resistance or should one follow the dictates of filial duty? Deliberation might fail to deliver a decisive verdict on this matter – and even if it does deliver a verdict, one need not act on it – but the mere fact that one has deliberated entails that one's agency has a different character from what it would have had, had one not deliberated.[6]

---

[6] Note that although only deliberative agency involves the active interrogation of the agent's reasons, both disinterested and deliberative agency exhibit what Fischer and Ravizza (1998) dub 'reasons-responsiveness'. Thanks to Neil Levy for this point.

Are Libet-actions automatic or willed? Some theorists have suggested that they lie towards the automatic end of the spectrum (Flanagan 1996). The idea, roughly, is that the decision to flex one's wrist *now* (and in such-and-such a way and at such-and-such a time) can be thought of as an automatic component of a complex, willed action. This complex action starts when the experimental procedure begins and one decides to flex one's wrist *at some point* in the next 30 seconds. Having consciously decided to comply with the experimental instruction, the subject offloads the execution of the motor response to automatic processes, with the result that the Libet-action proper is unconsciously initiated.

Although it is important to recognize that Libet-actions are embedded in a wider agentive context – a context that includes a conscious decision to produce an action of a certain type within a certain temporal window – I am not convinced that Libet-actions are best thought of as automatic. Unlike standard examples of automatic actions, subjects in Libet's experiments are explicitly required to attend to their own agency, and they do report that they decided (had an urge, intended) to produce the action immediately prior to doing so. Libet-actions may not be the 'ideal examples' of fully spontaneous agency that Libet himself takes them to be, but they do seem to be genuine instances of willed agency nonetheless. But although Libet-actions involve an act of will they do not involve deliberation – at least, not immediately prior to the action. In my terms, they are examples of disinterested agency, for the agent has no reason to flex their wrist at one particular time rather than the other, or to flex it in one way rather than another. Indeed, Libet-experiments are explicitly constructed so as to minimize the rational constraints under which the subject acts. We might think of Libet-actions as manifesting the liberty of indifference.[7]

With the foregoing in hand, let us return to the question of whether Libet-actions are paradigms of free will (as we intuitively conceive of it). Are disinterested actions our central exemplars of free will, or does that epithet belong to deliberative actions? Philosophers do not agree on the answer to this question, and the systemic research that would be required

[7] The free will sceptic might accept the foregoing, but suggest that neuroscientific models of agency derived from Libet's experiments have the potential to generalize to deliberative actions. Although it would be premature to dismiss this line of thought, the evidence to date suggests that the RP behaves very differently in the context of deliberative agency (Pockett and Purdy 2011).

in order to settle this dispute has not been carried out. That being said, my suspicion is that Roskies is right to identify the central or core cases of free will – at least, the kind of free will that is most intimately related to moral agency – with deliberation and rational reflection.

But even though Libet-actions might not be paradigms of free agency, it seems clear that they *do* fall within the scope of our pre-theoretical notion of free will. (Indeed, I suspect that common sense is inclined to regard even certain *automatic* actions as manifesting some degree of free will; we will return to this point.) As such, the free will sceptic is perfectly within his or her rights to claim that if Libet-actions – and indeed disinterested actions more generally – are not free then an important component of our common-sense conception of free will would be threatened. In sum, although (4) is unacceptable as stated, the sceptical argument is not thereby rendered impotent, for the question of whether Libet-actions manifest free will is itself an important one. Libet-actions might not qualify as ideal examples of free will, and they certainly do not provide us with the only form of agency that might be of interest to the neuroscience of free will, but they do provide the free will sceptic with a legitimate target.

4. The initiation of free actions

Let us turn now to the second premise of the sceptical argument: '(2) In order to exemplify free will an action must be initiated by a conscious decision'. We can think of (2) as the 'conceptual' step of the sceptical argument, for its plausibility turns chiefly on the contours of our everyday (or 'folk') notion of free will.

In order to determine whether (2) is plausible, we need to consider in what sense an event might be said to 'initiate' an action. I will work with a distinction between a strong sense of initiation and a weak sense of initiation. In the weak sense of the term, an event (ε) initiates an action (α) if and only if ε is the point of origin of α. The strong sense of initiation entails the weak notion, but adds the requirement that ε must be uncaused. This distinction between two notions of 'initiation' leads to two readings of (2):

WEAK: In order to exemplify free will an action must have its point of origin in a conscious decision.

STRONG: In order to exemplify free will an action must have as its point of origin a conscious decision which is itself uncaused.

Let us consider first what might be said on behalf of STRONG. This conception of free will characterizes decisions as 'unmoved movers' – they are the ultimate point of origin of action beyond which the causal chain cannot be traced. Should this constraint on free will be accepted?

Some incompatibilists might argue that it should. Incompatibilists hold that the truth of determinism – the thesis that a description of the current state of the world together with the laws of nature entails a description of all future states of the world – would rule out the possibility of free will. (Compatibilists, by contrast, hold that there is no incompossibility between free will and determinism.) Incompatibilists might be inclined to endorse STRONG on the grounds that conscious initiation is one way – indeed, perhaps the most straightforward way – in which indeterminism could enter into the structure of free agency. Compatibilists, on the other hand, are unlikely to be attracted to STRONG for they deny that free will requires the existence of uncaused causes. So, in order to determine whether STRONG should be accepted, we may need to first determine whether the notion of free will – that is, the *folk* notion of free will – is to be understood along compatibilist or incompatibilist lines.

We simply do not know the answer to this question. Systematic research into the structure and commitments of the folk or common-sense notion of free will has only just begun, and the results obtained thus far do not paint a clear picture. Some studies suggest that the folk are predominantly compatibilists, others suggest that the folk are predominantly incompatibilists.[8] One possibility is that 'the folk' do not share a single notion of free will, but that some of the folk are compatibilists and others are incompatibilists. Another possibility is that 'the' folk notion of free will contains both compatibilist and incompatibilist strands, each of which can be elicited depending on how the subject is probed (Nichols 2006). We do not as yet know enough about the contours of the common-sense notion of free will to decide between these possibilities. In light of this, any version of the sceptical argument that was committed to STRONG would be hostage to the results of future empirical inquiry.

Leaving the commitments of the folk notion of free will to one side, one might argue that STRONG is entailed by the conception of decisions that I introduced in Section 2, in which I said that the functional role of a decision is to settle what is unsettled. One might argue that this view entails

[8] For reviews of this literature, see Nahmias *et al.* (2005) and Nichols (2006).

that decisions cannot have fully sufficient causes, for any decision that had a fully sufficient cause would not need to settle anything – indeed, there would be nothing for it to settle.

Although superficially attractive, this line of thought fails to recognize that decisions settle *psychological* uncertainty (Holton 2006). An agent is required to make a decision only when their current psychological states fail to determine what they will do. However, an agent's behaviour could be psychologically 'unsettled' without being neurally 'settled'. Decisions lack fully sufficient psychological causes, but this does not entail that they (or their neural substrates) lack fully sufficient neural causes, and a decision can initiate an action even if it has a fully sufficient neural cause. So, STRONG is not entailed by the conception of decisions that we have employed.[9]

I have suggested that we have no good reason to embrace STRONG. What about WEAK? Should we require that free actions have their point of origin in a conscious decision?

The first comment to make is that the very thought that an action can always be traced back to a *single* point of origin is open to challenge. Rather than thinking of actions as originating with particular discrete events, we might do better to conceive of them as the outcome of multiple events and standing states, no single one of which qualifies as 'the' point of origin of the action. Just as the Nile has more than one tributary, so too many of our actions might result from multiple sources.

Secondly, to the extent that free actions can be traced back to a point of origin, it is by no means obvious that this point of origin must always be a conscious decision. Consider a thoughtless comment that is uttered on the spur of the moment and without forethought. Despite the fact that such an utterance is not consciously initiated, it would be very natural to hold the person who made it responsible for what they had said, and thus to assume that the notion of free will has some kind of grip in such contexts. But, the objection continues, if that is right, then (2) is too demanding, and freely willed actions need not be initiated by conscious decisions.[10]

---

[9] The only kind of determinism that might threaten the possibility of decision-making would be psychological determinism – that is, a determinism involving the agent's standing psychological states. Decisions settle what is unsettled, and if the agent's actions are fully caused by her standing psychological states then they would already be settled (whether or not the agent herself is aware of this fact). But Libet's experiments provide no support whatsoever for psychological determinism.

[10] For discussions of this issue, see Arpaly (2003), Levy and Bayne (2004), Smith (2005) and Sher (2009).

There are a number of points one might make in response to this objection. For one thing, the advocate of WEAK might deny that automatic actions are genuine examples of free will – or at least, that they are genuine examples of the kind of robust free will required for moral agency. Alternatively, they might allow that automatic actions can manifest free will in some sense, but only in virtue of the fact that they are suitably grounded in *prior* exercises of willed agency. (For example, one might take the thoughtless remark to manifest character traits that have been shaped by the agent's previous conscious decisions.) In light of this thought, we might replace WEAK with something akin to the following:

WEAK*: In order to exemplify free will an action must either have its point of origin in a conscious decision, or it must be appropriately grounded in a prior action that had its point of origin in a conscious decision.

Although this emendation is welcome, it doesn't really get to the heart of the matter. As we noted in Section 3, Libet-actions are not best categorized as automatic. So, although we may well need to modify WEAK in order to find a place for free will within the context of automatic agency, the crucial issue that confronts us in attempting to address the sceptical argument concerns the role played by conscious decisions in the context of *willed* actions. The central point here is this: Libet-actions, unlike automatic actions, are accompanied by the 'phenomenology of conscious initiation': the agent experiences themselves as deciding to act here-and-now, and – arguably – they experience this decision as the point of origin of the action in question.[11] At the centre of the sceptical argument is the thought that if Libet's RP data are correct then this experience of 'origination' must be an illusion, for Libet's data show that the action is already underway. We cannot, I suggest, attempt to preserve free will in the context of Libet-actions by attempting to ground such actions in prior acts of conscious initiation.

Where does this leave premise (2)? I have suggested that (2) can be understood in two quite different ways depending on the notion of

[11] See Horgan (2011) for a somewhat different view of the agentive phenomenology that accompanies Libet-actions. Horgan acknowledges that one would experience oneself as beginning to actively undertake an action at some specific moment in time, but he denies that this phenomenology would involve the sense that one's behaviour was caused by one's mental states. Instead, he suggests, one would experience oneself as the *author* of one's actions, where this phenomenology of authorship should not be thought to include any experience as of mental state causation. Although the issues raised by Horgan's discussion are important I lack the space to do them justice here.

initiation that is deployed. On a strong reading of the term, (2) requires that free actions must begin with uncaused causes. Although certain incompatibilist conceptions of free will might be committed to this view, it is an open question how central this view is to the common-sense notion of free will. A weak reading of 'initiate', by contrast, requires only that free actions have their point of origin in a conscious decision. Although even this claim is unacceptable as it stands, there is something to it, for I have suggested that agents in the Libet paradigm do experience their decisions as the point of origin of those simple motor behaviours that they produce in such contexts. The question is whether this experience is at odds with the neural data that Libet obtained. To answer that question we need to consider the first premise of the sceptical argument.

5. Conscious decisions and the readiness potential

The first premise of the sceptical argument is: '(1) The actions studied in the Libet paradigm are not initiated by conscious decisions but are instead initiated by the RP'.

There are a number of ways in which one might attempt to put pressure on this premise. I will begin by considering the possibility that (1) operates with a false contrast between the agent's decision and the RP. The idea here is that the RP and the agent's decision might be the same event – or at least, that the RP might be the neural basis of the agent's decision.[12] If this could be established, then we might conclude that the agent's actions are initiated by both the RP *and* their decision.

On the face of things this proposal might appear to be a non-starter. After all, isn't there a gap of around 300–350ms between the RP and the W judgements that subjects make? Given this gap, how could it be an open epistemic possibility that the RP is the same event as the agent's decision? In order to answer this question, we must distinguish two ways in which experiences can be associated with temporal properties. On the one hand, experiences can *represent* temporal properties. For example, one might

---

[12] There are various conceptions of the relationship between mental events and neural events. One might take the two kinds of events to be merely correlated with each other, one might take mental events to supervene on or be realized by neural events, or one might take mental events to be identical to neural events. I use 'basis' here in a way that is neutral between these various accounts.

experience a flash of light as occurring before an explosion. This is the temporal structure of the *contents* of experience. At the same time experiences themselves have temporal properties. The experience of the flash and the experience of the explosion will have certain locations in objective time. We might call these properties the *vehicular* properties of experience.

Libet's experiments are concerned with both forms of temporal structure. If the RP data tell us anything about the temporal properties of mental states, then they tell us about their properties considered as vehicles. They tell us when those experiences occur. However, the W data is best understood in terms of the contents of the agent's experiences. The W judgement represents the agent's opinions as to when their decision to flex their wrist occurred. However, the W judgement itself might occur at some other time. The agent might (or might not) have access to the temporal location of the W judgement, but if so, they don't have introspective access to it simply in virtue of making a W judgement. This is because introspection provides us with access only to the *contents* of experience (see, e.g., Tye 2002). So, we need to understand W in terms of the subject's representation of the time of their decision.

The foregoing provides us with a way in which we can see the assumption that is built into (1) might be false, for there is no a priori reason to assume that the temporal location that is represented in the content of an experience must be identical to the temporal location of the experience itself – that is, of its vehicle. As Daniel Dennett and Marcel Kinsbourne (1992) pointed out, just as there is no a priori requirement that the brain employ spatial properties to represent spatial properties, so too there is no a priori requirement that it use temporal properties to represent temporal properties. With this point in hand, Dennett (1991) suggests that the appearance of a gap between the RP and the W judgement results from a failure to appreciate the fact that we are dealing with two kinds of temporal properties. Perhaps, he suggests, the RP is the neural basis of a judgement whose content is expressed in the W judgement: although this mental state itself occurs 550ms prior to the action, it is represented as occurring only 200ms prior to the action.

There are two points to make about this proposal. First, it sits uneasily with a certain conception of the content of decisions. Arguably, the contents of W judgements are token-reflexive. In other words, the referent of 'now' in the decision to flex one's wrist 'now' is simply the location of that decision itself. So, if the RP is the neural basis of that decision, then one has in fact decided to flex one's wrist 550ms prior to the action, rather than only

200ms prior to the action as the subject reports. Secondly, even if this proposal shows that there is no incoherence in the thought that the RP could be the neural basis of the agent's decision, we have no evidence that the RP actually *does* form the neural basis of the agent's decision. At present, this proposal represents nothing more than an intriguing account of how the agent's decisions might be related to the RP.[13]

David Rosenthal has attempted to undermine (1) from a slightly different angle. Drawing on his higher-order thought account of consciousness, Rosenthal (2002) argues that Libet's results are not at all surprising. According to the higher-order account, a mental state is conscious only in virtue of the fact that it is the intentional target (or object) of another mental state. On this view, decisions and intentions are conscious in virtue of the fact that the agent is conscious *of* them, where this 'consciousness of' involves a higher-order mental state which represents that one has decided or intended to do such-and-such. Given that this higher-order state is subsequent to the decision itself, it is only to be expected – Rosenthal argues – that W judgements lag behind the RP. Although Rosenthal draws heavily on his higher-order theory of consciousness, his proposal is actually independent of any general commitment to the higher-order treatment of consciousness. One could hold that although certain types of mental states are conscious independently of their being monitored by other mental states, as it happens we are conscious of our decisions and intentions only by monitoring them in a certain way.

What should we make of Rosenthal's account? Some theorists will dismiss it on the grounds that intentions and decisions are intrinsically conscious. On this view, the suggestion that decisions or intentions might be unconscious is no more coherent than is the notion that a pain or an itch might be unconscious – decisions and intentions *just are* events in the stream of consciousness.

I don't think that this objection is at all plausible as far as *intentions* are concerned. Consider first distal intentions – intentions to do something in the medium to long term. This morning I formed the intention to spend Christmas in Switzerland, but having formed that intention I promptly put the matter out of my mind and was not conscious of it until now.

---

[13] Furthermore, the fact that the correlation between W judgements and LRP activity is more robust than is the correlation between W judgement and RP activity (see below, p. 40) suggests that LRP activity is more likely than RP activity to function as the neural basis of the agent's decision.

Undeterred, the advocate of the objection might grant that distal intentions need not be conscious, but they might insist that proximal intentions – intentions to do something in the immediate future – must be conscious. But even that claim seems implausible. Consider what it's like to carry out a routine and over-learned action, such as climbing steps or changing gear, whilst lost in thought. In such cases, one's behaviour is guided by intentions (to move this object over here; to move that object over there) of which one may be unaware. There is little reason to insist that intentions must be conscious even when they are active in governing one's behaviour (Mele 2009).

Decisions, on the other hand, might be a different matter. Certainly our intuitive conception of decisions seems to be strongly wedded to the assumption that decisions are acts that one carries out consciously; they are not mental events of which one may (or may not) become conscious. Of course, one might argue that although folk psychology does assume that decisions are intrinsically conscious, this assumption is up for grabs in the sense that we can make sense of the thought that certain decisions might be unconscious. I'm not so sure that we can. This isn't to say that there aren't legitimate uses of the term 'decision' according to which decisions can be unconscious. Consider, for example, the fact that cognitive neuroscientists describe the visual system as 'deciding' how to categorize a perceptual stimulus. However, one might want to draw a fairly clear distinction between such 'sub-personal' decisions and the kind of personal-level decisions with which we are interested here. Certainly there is some reason to demand that the kind of decisions connected with agent-level responsibility must be conscious. At the very least, a certain amount of suspicion surrounds the notion of an unconscious decision.[14]

Thus far we have considered two ways in which one might attempt to 'identify' the agent's decision with the RP, and I have suggested that there are objections to both accounts. However, these objections are not obviously decisive, and it is possible that one (or both) of these accounts can be patched up. Even so, the free will sceptic might argue, neither account would really provide a vindication of free will. 'The reason for this', the sceptic might say, 'is that it is not the agent's conscious decision *as such* which leads them to act but rather the neural activity which underlies that decision. The mental property *per se* is not causally

---

[14] If we do allow unconscious decisions then we will need non-introspective tools for identifying such states, and it is not entirely clear what such tools might be available.

efficacious; instead, the only causally efficacious properties here are neural properties. The decision is a "free-rider", and hence there is no room for genuine free will.'[15]

This line of thought is one manifestation of what is known as the 'causal exclusion problem'. The worry is that the causal efficacy of mental properties is 'excluded' or 'screened off' by the causal efficacy of the neural properties underpinning them. This problem has generated a small mountain of literature, and there is as yet no consensus on how – if at all – it might be solved (see, e.g., Bennett 2008; Kallestrup 2006; Kim 1993). However, this is not a literature that we need engage with here, for the issues raised by the exclusion problem are much more general than those which are raised by Libet's data. All that is required in order to provide some intuitive motivation for the exclusion problem is the assumption that the agent's mental events have a physical basis, and most philosophers of mind endorse that assumption independently of any appeal to Libet's experiments. Any comprehensive response to free will scepticism must include a solution to the causal exclusion problem, but it is not incumbent on us to provide that solution here.

At the outset of this section I stated that there are two ways in which one might attempt to resist (1). Thus far we have explored the idea that the RP might function as the neural basis of the agent's decision, and hence that both the RP and the agent's decision might be said to initiate the action in virtue of this single event. However, let us assume that the RP is not the neural basis of the agent's decision, and that the RP occurs before the agent decides what to do. Would this fact show that the agent's action is initiated by the RP rather than by his or her decision?

Not necessarily. To see why, we need to return to the question of what it is for an event to initiate an action. I suggested in Section 4 that an event (ε) initiates an action (α) only if ε functions as α's point of origin. The notion of a 'point of origin' can be variously understood, but let us say that ε functions as α's point of origin only if there is a robust correlation between ε-type events and α-type events, such that in normal contexts there is a high probability that an ε-type event will be followed by an α-type event. (The notion of origination requires more than this, but it is implausible to suppose that it requires less than this.) So, if the RP is the origin of the

---

[15] This kind of worry is suggested by Haggard's comment that 'Although consciousness may be part of brain activity, consciousness cannot cause brain activity, nor can it cause actions' (this volume: p. 18).

agent's action, then we ought to expect RP events to be 'immediately' followed by the appropriate action, unless something unusual happens (such as the person being struck by lightning). Is this the case?

As Mele (2009) and Roskies (2011) have observed, we simply do not know. Our ignorance on this point derives from the fact that the RP is measured by a process known as 'back-averaging'. Because the RP on any one trial is obscured by neural noise, what is presented as 'the RP data' is determined by averaging the data collected on a large number of trials. In order to compute this average, the EEG recordings on different trials need to be aligned, and this requires some fixed point – such as the onset of muscle activity or some other observable behaviour on the part of the subject – that can be identified across trials. Any RPs that are not followed by an action simply won't be measured, and so we don't know how robust the correlation between the RP and Libet-actions is.[16]

We do, however, have indirect reasons to think that the relation between the RP and subsequent action may not be as tight as that which would be required in order for the RP to function as the point of origin of the action. First, we know that the nature of the experimental context can significantly affect both the temporal properties and the strength of the RP signal. Subjects who are highly motivated to perform the task produce a large RP, whereas the RP almost disappears in subjects who have lost interest in the task (McCallum 1988; Deecke *et al.* 1973; Rigoni *et al.* 2011). Secondly, it is possible to make willed responses to stimuli in very much less than 550ms, which indicates that the RP is not 'the' point of origin even where it occurs. Thirdly, another neural event – the *lateralized* readiness potential (LRP) – appears to be more strongly coupled to agency than the (generalized) RP is. Whereas the (generalized) RP is symmetrically distributed over both hemispheres, the LRP is restricted to the hemisphere contralateral to the hand that is moved. (In other words, one will see a left-hemisphere LRP for a right-handed action and a right-hemisphere LRP for left-handed actions.) As Haggard points out (this volume), the evidence indicates that the LRP is more robustly correlated with the subsequent action than the

---

[16] Some commentators have also worried that because the Libet experiments involve averaging across a number of trials, certain aspects of the data might be statistical illusions. In other words, features of the relationship between (say) the RP and the W judgement might characterize the averaged data even though they do not characterize any of the individual trials that contribute to that grouped data. For discussion of this issue, see Roskies (2011) and Trevena and Miller (2002).

RP is (see Haggard and Eimer 1999). However, the LRP is also very tightly coupled to the W judgements that subjects make, and thus a version of the Libet argument in which (1) is replaced with a corresponding claim about the LRP does not possess even the surface plausibility that (1) does.

Taken together, these three points suggest that the RP is unlikely to qualify as 'the' point of origin of the action. If the RP has a psychological interpretation – and it is far from clear that it does – then we should perhaps think of it as the neural realization of quite general cognitive and motivational features that contribute to agency. We might think of the RP as the neural basis of an 'urge' or 'inclination' to act, rather than as the neural basis of the decision to act now (Gomes 1999; Mele 2009). It is one of the many tributaries that contribute to the formation of the action, but it is not 'the origin' of the action in any intuitive sense of that term.[17]

Let me recap the argument of this section. I began by considering responses to (1) which argue that it is possible that Libet-actions could be initiated by both the agent's decision and by the RP, for it might turn out that the RP is the neural basis of the agent's decision. I suggested that although this line of thought should not be dismissed, various considerations weigh against it. I then examined a second, and more plausible, objection to (1) – namely, that the very means of measuring the RP prevents us from determining the robustness of the correlation between it and whether the agent acts. Finally, I argued that various indirect considerations suggest that the correlation between the RP and Libet-actions is not robust enough for us to be justified in describing the RP as the origin of the Libet-action. Instead, the RP may be no more than one of several elements, each of which contributes to the production of the Libet-action.

6. Conclusion

In this chapter I have examined the standard – and arguably most powerful – version of the argument for free will scepticism based on the results of Libet's experiments. I began with the fourth premise of the sceptical

---

[17] One might argue for a similar account of the data reported in Soon *et al.* (2008), in which the researchers were able to predict which of two decisions agents made up to 10s before they acted by measuring activity in prefrontal and parietal cortex. This neural activity clearly contributes to the agent's decision, but it is far from clear that it 'initiates' the action.

argument, and the question of whether Libet-actions qualify as a legitimate target for the scientific investigation of free will. Even though it is doubtful that they are the ideal exemplars of free will that Libet takes them to be, they do fall within the scope of those actions that we intuitively regard as manifesting free will. The discovery that Libet-actions are not freely performed would not itself show that *none* of our actions are freely performed, but it would go some way towards vindicating free will scepticism.

I then turned to the second (or 'conceptual') premise of the sceptical argument, which claims that freely willed actions must be initiated by conscious decisions. I began by distinguishing a 'strong' notion of conscious initiation from a 'weak' notion. The strong notion requires that initiating decisions are uncaused, whereas the weak notion imposes no such requirement, although it does require that initiating decisions have no fully sufficient psychological causes. I suggested that although the precise content of the folk notion of free will is open to debate, it is doubtful whether the folk are committed to the claim that freely willed actions must be consciously initiated in the strong sense of that notion. I also noted that there are questions concerning both the requirement that freely willed actions must be consciously initiated, and the requirement that free actions must be directly initiated in an act of will.

Finally, in Section 5, I turned to the first (or 'empirical') premise of the sceptical argument: the claim that Libet-actions are not initiated by conscious decisions but are instead initiated by the RP. We saw that there are two ways in which one might put pressure on this premise: by arguing that it might be possible to identify the RP with the neural basis of the agent's decision, and by arguing that the RP is merely a contributing factor for the relevant action rather than its point of initiation. I suggested that although the first response to (1) is problematic, there is much to be said in favour of the second line of response. All in all, rumours of the 'death' of free will are, I have argued, greatly exaggerated.

What *would* constitute a neutrally based objection to free will? This is a surprisingly difficult question to answer, and a lot will depend on just what the ordinary, intuitive conception of free will is committed to. That said, here is one line of evidence that *might* place our intuitive commitment to free will under strain. Suppose that one found evidence of a neural state that one knew functioned as the neural realization of a psychological state (ϕ) where ϕ occurs immediately prior to the agent's decision. Further, suppose that one had independent reason to think that ϕ was a fully sufficient cause of the agent's decision. Such a discovery, I suggest, would be at odds with the agent's sense of herself as deciding what to do – as 'making up her mind'. What one would have discovered is that the agent was not in the state of psychological uncertainty that she took herself to be in. She experienced herself as 'making up her mind', but in fact her mind was already 'made up'. For what it's worth, my hunch is that the sciences of human agency are exceedingly unlikely to provide us with evidence along these lines, but that's a matter for another occasion.

Although my central focus in this chapter has been with the sceptical challenge to free will, I do not want to give the impression that this challenge constitutes the only – or even the most important – point of contact between the neuroscience of agency and questions of free will. Rather than asking whether the sciences of human agency undermine our commitment to free will, we might instead look to them for insights into the nature of free will. What is it about human cognitive architecture that provides us with the capacity for free agency? How can the domain of human freedom be expanded? How can impediments to the exercise of free will be removed? With few exceptions (see, e.g., Holton 2009; Roskies 2010), philosophical engagement with the sciences of agency has been dominated by attempts to address the sceptical challenge. That focus has been understandable, but perhaps it is time for us to consider how neuroscience might enrich our understanding of free and autonomous agency.

References

Arpaly, N. (2003) *Unprincipled Virtue*, New York: Oxford University Press.

Banks, W.P. and Pockett, S. (2007) Benjamin Libet's work on the neuroscience of free will. In M. Velmans and S. Schneider (eds) *The Blackwell Companion to Consciousness*, Oxford: Blackwell, pp. 657–70.

Banks, W.P. and Isham, E.A. (2011) Do we really know what we are doing? Implications of reported time of decision for theories of volition. In W. Sinnott-Armstrong and L. Nadel (eds) *Conscious Will and Responsibility*, New York: Oxford University Press, pp. 47–60.

Bennett, K. (2008) Exclusion again. In J. Hohwy and J. Kallestrup (eds) *Being Reduced*, Oxford: Oxford University Press, pp. 280–306.

Deecke, L., Becker, W., Grözinger, B., Scheid, P. and Kornhuber, H.H. (1973) Human brain potentials preceding voluntary limb movements. In W.C. McCallum and J.R. Knott (eds) *Electroencephalography and Clinical Neurophysiological Supplement: Event-related Slow Potentials of the Brain: Their Relations to Behavior* (Vol. 33), Amsterdam: Elsevier, pp. 87–94.

Dennett, D. (1991) *Consciousness Explained*, Boston: Little, Brown and Company.

Dennett, D. and Kinsbourne, M. (1992) Time and the observer, *Behavioral and Brain Sciences*, 15: 183–247.

Fischer, J.M. and Ravizza, M. (1998) *Responsibility and Control: A Theory of Moral Responsibility*, New York: Cambridge University Press.

Flanagan, O. (1996) Neuroscience, agency, and the meaning of life. In *Self-Expressions*, Oxford: Oxford University Press.

Gomes, G. (1999) Volition and the readiness potential, *Journal of Consciousness Studies*, 6(8/9): 59–76.

Haggard, P. (2006) Conscious intention and the sense of agency. In N. Sebanz and W. Prinz (eds) *Disorders of Volition*, Cambridge, MA: MIT Press, pp. 69–86.

Haggard, P. (2008) Human volition: towards a neuroscience of will, *Nature Reviews Neuroscience*, 9: 934–46.

Haggard, P. and Eimer, M. (1999) On the relation between brain potentials and the awareness of voluntary movements, *Experimental Brain Research*, 126: 128–33.

Haggard, P., Newman, C. and Magno, E. (1999) On the perceived time of voluntary actions, *British Journal of Psychology*, 90: 291–303.

Hallett, M. (2007) Volitional control of movement: the physiology of free will, *Clinical Neurophysiology*, 118: 1179–92.

Holton, R. (2006) The act of choice, *Philosophers' Imprint*, 6(3): 1–15.

Holton, R. (2009) *Willing, Wanting, Waiting*, Oxford: Oxford University Press.

Horgan, T. (2011) The phenomenology of agency and the Libet results. In W. Sinnott-Armstrong and L. Nadel (eds) *Conscious Will and Responsibility*, New York: Oxford University Press, pp. 159–72.

Horgan, T., Tienson, J. and Graham, G. (2003) The phenomenology of first-person agency. In S. Walter and H.-D. Heckmann (eds) *Physicalism and Mental Causation: The Metaphysics of Mind and Action*, Exeter: Imprint Academic, pp. 323–40.

Kallestrup, J. (2006) The causal exclusion argument, *Philosophical Studies*, 131(2): 459–85.

Keller, I. and Heckhausen, H. (1990) Readiness potentials preceding spontaneous motor acts: voluntary vs. involuntary control, *Electroencephalography and Clinical Neurophysiology*, 76: 351–61.

Kim, J. (1993) The non-reductivist's troubles with mental causation. In J. Heil and A. Mele (eds) *Mental Causation*, Oxford: Clarendon, pp. 189–210.

Lau, H.C., Rogers, R.D., Haggard, P. and Passingham, R.E. (2004) Attention to intention, *Science*, 303(20): 1208–10.

Levy, N. (2005) Libet's impossible demand, *Journal of Consciousness Studies*, 12: 67–76.

Levy, N. and Bayne, T. (2004) Doing without deliberation: automatism, automaticity, and moral accountability, *International Review of Psychiatry*, 16(4): 209–15.

Libet, B. (1985) Unconscious cerebral initiative and the role of conscious will in voluntary action, *Behavioral and Brain Sciences*, 8: 529–66.

Libet, B., Wright, E.W. and Gleason, C.A. (1982) Readiness potentials preceding unrestricted spontaneous preplanned voluntary acts, *Electroencephalographic and Clinical Neurophysiology*, 54: 322–5.

Libet, B., Gleason, C.A., Wright, E.W. and Pearl, D. (1983) Time of unconscious intention to act in relation to onset of cerebral activity (readiness-potential), *Brain*, 106: 623–42.

McCallum, W.C. (1988) Potentials related to expectancy, preparation and motor activity. In T.W. Picton (ed.) *EEG Handbook Volume 3: Human Event-Related Potentials*, Amsterdam: Elsevier, pp. 427–534.

Mele, A. (2009) *Effective Intentions: The Power of Conscious Will*, New York: Oxford University Press.

Miller, J.O., Vieweg, P., Kruize, N. and McLea, B. (2010) Subjective reports of stimulus, response, and decision times in speeded tasks: how accurate are decision time reports? *Consciousness and Cognition*, 19: 1013–36.

Nahmias, E., Morris, S.G., Nadelhoffer, T. and Turner, J. (2005) Surveying freedom: folk intuitions about free will and responsibility, *Philosophical Psychology*, 18(5): 561–84.

Nichols, S. (2006) Folk intuitions on free will, *Journal of Cognition and Culture*, 6(1/2): 57–85.

Pockett, S. (2004) Does consciousness cause behaviour? *Journal of Consciousness Studies*, 11(2): 23–40.

Pockett, S. and Purdy, S. (2011) Are voluntary movements initiated preconsciously? The relationships between readiness potentials, urges, and decisions. In W. Sinnott-Armstrong and L. Nadel (eds) *Conscious Will and Responsibility*, New York: Oxford University Press, pp. 34–46.

Rigoni, D., Kühn, S., Sartori, G. and Brass, M. (2011) Inducing disbelief in free will alters brain correlates of preconscious motor preparation, *Psychological Science*, 22(5): 613–18.

Roediger, H.K., Goode, M.K. and Zaromb, F.M. (2008) Free will and the control of action. In J. Baer, J.C. Kaufman and R.F. Baumeister (eds) *Are We Free?* Oxford: Oxford University Press, pp. 205–25.

Rosenthal, D. (2002) The timing of conscious states, *Consciousness and Cognition*, 11: 215–20.

Roskies, A. (2010) How does neuroscience affect our conception of volition? *Annual Review of Neuroscience*, 33: 109–30.

Roskies, A. (2011) Why Libet's studies don't pose a threat to free will. In W. Sinnott-Armstrong and L. Nadel (eds) *Conscious Will and Responsibility*, New York: Oxford University Press, pp. 11–22.

Sher, G. (2009) *Who Knew? Responsibility Without Awareness*, Oxford: Oxford University Press.

Sinnott-Armstrong, W. and Nadel, L. (2011) *Conscious Will and Responsibility*, New York: Oxford University Press.

Smith, A. (2005) Responsibility for attitudes: activity and passivity in mental life, *Ethics*, 115: 236–71.

Soon, C.S., Brass, M., Heinze, H.-J. and Haynes, J.-D. (2008) Unconscious determinants of free decisions in the human brain, *Nature Neuroscience*, 11(5): 543–5.

Spence, S. (2009) *The Actor's Brain: Exploring the Cognitive Neuroscience of Free Will*, New York: Oxford University Press.

Trevena, J.A. and Miller, J. (2002) Cortical movement preparation before and after a conscious decision to move, *Consciousness and Cognition*, 11: 162–90.

Tye, M. (2002) Representationalism and the transparency of experience, *Noûs*, 36: 137–51.

Wegner, D.M. (2002) *The Illusion of Conscious Will*, Cambridge, MA: MIT Press.

3
Physicalism and the determination of action

FRANK JACKSON

1. What I'll do

There is no single version of physicalism. There is no single argument for physicalism. There is, accordingly, no standard answer concerning the implications of physicalism for the causation of human action by mental states. Of necessity, I will be highly selective. I will start by describing the version of physicalism I favour and saying why I favour it. We will then discuss what it implies about the connection between subjects' mental states and what they do, and thereby for the determination and pre-dictability of our actions. This serves as a precursor for a short discussion of the implications of physicalism for the possibility of free action. Near the end I will say something about a version of physicalism that I don't favour but which many like. This version is commonly thought to have very different implications for the determination and prediction of action from the version of physicalism that I like. We will see that this is indeed the case but not in ways that help much when we are worrying about the impli-cations of physicalism for the possibility of free action.

2. The preferred version of physicalism

Give a stone a kick and not much happens. Give a person a kick and quite a lot may happen. Present an apple with a quadratic equation and not much happens. Present a person with a quadratic equation and you may get the answer. Present a person with a set of house plans and you may end up with a house, no point though in doing this with a tree.

These commonplaces remind us that persons determine functions from inputs to outputs of a highly sophisticated kind. What is more, we know these functions connect closely with a person's mental states. If someone is required to cross a minefield, their response tells you very quickly where they believe the mines are located. That is, the function you observe from minefield to response tells you something about their mental states, namely, where they believe the mines are located. Likewise, we know that the function from house plans to built houses goes via what the potential builder sees and wants. Showing plans to someone who can't see them or doesn't want to build the house won't get you very far.

What's the take-home message? That mental states play functional roles. This is hardly news. It is, surely, something the folk know. It is not something discovered by high-powered cognitive science. The folk have known since the dawn of time about the *functional* impairments that go along with failing eyesight and hearing, for example, and that these impairments relate directly to the nature of their visual and auditory experiences.

There is a simple way to go from these commonplaces to a version of physicalism. Here is how it goes:

> Premise 1
> Mental state so and so = the state that plays role such and such. (Commonplace we have just been talking about)
>
> Premise 2
> The state that plays role such and such = brain state whatever, having the kinds of properties that figure in current and future neuroscience. (Empirical discovery)
>
> Conclusion
> Mental state so and so = brain state whatever. (By the transitivity of identity)

Some say the first premise, suitably expanded and detailed, is a conceptual truth of a kind with 'Poisons make people sick'. The idea is that for each mental state there is a *definitive* functional role. This issue doesn't matter for our purposes right now (although we return to the question later). What matters for now is that something like the first premise is true for the mental states we will be concerned with, among which are our intentions and our decisions to do this or that.

Some say the conclusion isn't really an identity; it should be read as a claim about constitution, analogous to the relation between a table and the molecules that compose it. Mental states are constituted by, while not being identical with, brain states. This issue is by the way for us here.

Some say the conclusion is a contingent truth; others say that – for suitable values of 'such and such' and 'whatever' – it is a necessary a-posteriori truth. We will return to this issue later.

What is important at this stage is the commitment to mental states playing causal roles. The functional roles we are talking about require that mental states play causal roles. They may require more than this, and what that more might be is contentious, but they require at least that mental states play causal roles. This means that Richard Swinburne (this volume) and I agree that epiphenomenalism is false. We are at one in rejecting the idea that mental states do nothing.

3. Mental states directed at mental states

We are now in a position to say something quickly about the debate initiated by the work of Benjamin Libet; the debate that in part prompted the symposium that this volume arises from. (For details of Libet's results, see Haggard, this volume.)

Among the functional roles played by mental states are roles directed towards mental states themselves. Not all of the functional roles concern how we relate to our surroundings. For example, perceptual experience is a putative response to the environment, a response that represents that things are thus and so, whereas the belief that one has an experience that represents that things are thus and so is a response to the perceptual experience itself. Similarly, although the job of implementing one's intention to obey someone is to cause behaviour in conformity with that intention, the job of one's *awareness* of implementing one's intention to obey someone is to let one know that you are indeed implementing the intention – something it can be very useful to know. Likewise, the functional role of the belief that it is raining is to record the putative information that it is raining and to lead to behaviour that tends to serve one's desires if that belief is true – taking an umbrella before going out, say. But that isn't the role of the awareness that one believes that it is raining. The role of the awareness that one believes that it is raining is, for example, to make it explicit to one what one believes and to generate, in animals with language, sentences like 'It is raining'.

This means – at least for anyone who likes the functionalist approach to mental states – that whenever we consider how best to interpret the results of experiments in psychology that draw heavily on subjects' own reports of, or judgements about, the mental states they are in, including, as it might be, the time at which the mental states occur, we need to be alert to the possibility that what's being reported is a property of the awareness of the mental state rather than the mental state putatively under investigation.

Now what Libet did was to devise an ingenious way of timing when subjects take it that they are making a certain decision. He found that the time at which they take it that they are making the decision is somewhat later than the time at which the relevant changes in the brain – those that lead to behaviour that implements the decision – commence. But this doesn't tell us that the decision *itself* takes place somewhat later, a result which would raise awkward questions about the causal role of the mental state of deciding to act. It tells us instead something about subjects' awareness of the time at which they implement a decision.[1]

4. Why believe the preferred version of physicalism?

The argument set out earlier is valid. We have seen why we should accept the first premise. The question that remains, accordingly, is the case for the second premise. This has been, as one would expect, the subject of much debate. (See any recent text in the philosophy of mind.) But, in a nutshell, the case rests on neuroscientific optimism combined with a general hostility to the idea of outside causal influences on the physical world.

There are lots of things we don't know about how our brains work but there doesn't seem any particular reason to hold that we will find some kind of gap in the causal story about how stimuli at our sense organs lead to motor responses via happenings in the brain, a gap suggesting that some kind of causal influence from outside the physical world is at work in the brain, or at some point in the causal path from environmental input to behavioural response.

The last paragraph is essentially a reminder of the line of thought that is often encapsulated by saying that the physical world is causally closed. Although physicalists like myself are supporters of the thesis that the physical world is causally closed – read in a way that takes account of indeterminism – there is a well-known problem in saying what 'physical'

---

[1] Which plausibly isn't the time of the awareness *per se* but the time at which the awareness represents the deciding to be taking place. Thanks here to Nic Shea.

means here. I'll say something about this later. For now, we can think in terms of the kinds of properties neuroscientists appeal to when they explain the way the impact of the world on our sense organs translates into motor responses. This means that the core idea behind the second premise is that when neuroscientists detail and expand our understanding of the causal paths from environmental impacts on our peripheries to happenings inside us, especially happenings in our brains, and the pathways to and from our brains, and from happenings elsewhere inside us, to behavioural responses, they won't find causal transitions that suggest the presence of outside causal influences.

Two more points to note about the second premise. First, I talked of the case for it resting on neuroscientific optimism. This suggests that anyone who wishes to resist the kind of physicalism I like only needs to be a bit of a pessimist. An easy thing to be, one might think. However it isn't that simple. We want the words that come out of our mouths, the marks we make on paper and what we do to keyboards to carry information about our mental states. If that were not true, my diary entries for, say, 2001 would not be a good source of information about what I thought and felt back then, and emails from your friends would not be good sources of information about what they are thinking and feeling. Anyone who wants to hold that a critical part of our mental nature lies in properties additional to those posited in neuroscience and the functional roles they fill, must hold that, somehow or other, these extra properties stand in information-preserving causal relationships to the words we use to talk about them, the various marks that appear on paper and what we do to keyboards. They need to be optimistic – implausibly optimistic, say I – about how the laws that govern the connections between the extra properties and what happens in our brains ensure that the distal happenings at our motor extremities carry information about the extra properties.

Secondly, there is a feature of our version of physicalism that some philosophers worry about. (My sense is that philosophers worry about it more than brain scientists.) Our version is a version of internalism about mental states; it affirms that they are located inside our heads, because that is, as a matter of fact, where the crucial functional roles are played.[2] We should, that is, locate them inside our heads for the same reason we should locate the memory processing in an iMac somewhere behind the screen.

---

[2] The roles themselves, however, concern in part how things are outside the subject. Part of what makes some state a desire for coffee is its connection with coffee.

The worry is that the states we are talking about are types or kinds of state, and types and kinds *as such* aren't located. (This seems, for example, to be Williamson's concern (2009: 331).) They are abstract entities. It is their instances – the things that are of the relevant type or kind – that are located. This might suggest that the only sensible thing to mean by the location of a mental state is the location of someone who is in that state, and we aren't located inside our heads.

There is nothing to worry about here, or so it seems to me. There are many states – many kinds of states – that a largish physical structure is in by virtue of some part of it being a certain way. A computer stores a memory in virtue of a part of it being the right way; a person has cancer in virtue of a part of that person being a bad way; someone has arthritis in virtue of a part of a joint being inflamed; and so it goes. In cases like these, although the largish thing is in the state, the state itself is located where the part of the big thing that is the relevant way is located.

You now have before you the version of physicalism I like; it is a version of the type-type mind-brain identity theory (see, e.g., Armstrong 1968). What does this version of physicalism say about the determination of action and the predictability of action, the questions that relate most directly to the topic of this volume?

The answers depend on two matters: how we should think about the first premise of the argument we started with, and how we should respond to what physics tells us about determinism.

5. The functionalist premise

There are broadly three ways we can think of the first premise: as a bit of common sense about mental states but nothing more than that; as something that, with suitable elaboration, can be turned into a conceptual truth; or as something that, with suitable elaboration, can be turned into an account of how we reference-fix on mental states.

The first downplays the importance of the functional roles played by mental states. I think it goes too far. People who suffer strokes want above all else to get function back. I think they would be surprised to be told that the connection between getting function back and getting, say, the ability to see things off to their left was purely accidental. Or consider what is involved in remembering something. The role, the functional role, of remembering in delivering information about the past seems much more than a well-known but accidental fact about remembering. Again, the observation that belief is a state designed to fit the world, and the observation that desire is a state designed to make the world fit it, seem to be much more than known truths about belief and desire. They seem to go to the heart of what belief and desire are. Finally, is it nothing more than an interesting truth that people in pain tend to behave in ways that they believe will minimize their pain?

Suppose then that the functional roles of mental states are more than interesting facts about them. One response, as we observed earlier, is to think of being in a mental state of such and such a kind as like being poisoned: something definable in terms of a certain functional role. In the case of being poisoned, we can state the functional role. In the case of any given mental state we cannot. We can give rough indications – our talk above of belief being a state designed to fit the world is an example, as is our talk earlier of perceptual experience being a putative response to the environment that represents that things are thus and so – but a statement with all the bells and whistles in place will be beyond us. Here is the point where those who like this approach, and I'm an example, talk about implicit theories. They say that mental states are defined by implicitly grasped functional roles.

The other response is to think of the functional roles played by mental states as reference fixers. It is obvious that we have a pretty good grasp of the functional roles played by various mental states. That is evident from the way we are able to explain and predict people's behaviour in terms of their mental states, and the way we use how they respond to various situations to reach reliable judgements about their mental states. But we need not think of these remarks as support for the existence of some sort of conceptual connection between mental states and functional roles. Maybe the functional roles serve to fix the reference of the mental state terms. On this view, the essential nature of any given mental state is the brain state it is identical with, much as, according to many, the essential nature of water is H_2O. But our knowledge that certain functional roles are occupied preceded our knowledge of the brain. Something about the way we and our fellows interacted with the world led us to hypothesize that, inside each of us, there are, for example, states that carry information about the world and which interact with each other in systematic ways. How else are we able to avoid the holes in the ground and the tigers? And how else can we explain the fact that, over time, people get better at avoiding the holes in the ground and the tigers – obviously some sort of useful information processing and updating is going on inside us. Accordingly, we should, says the reference-fixing view, think of the functional roles as alerting us to the existence of internal states that are connected one to another, and to the world around us, in various ways. These internal states are our mental states but their essential nature *qua* mental states isn't given by their functional roles. That's a contingent feature of them that led us to hypothesize their existence in the first place. Their essential natures are instead given by their a-posteriori-determined neurological natures.

We have here two very different ways of thinking about the identities between mental states and brain states that are the conclusion of the two-premise argument for physicalism we set out earlier. If the connection between functional role and mental state is conceptual, the mind-brain identities will be contingent a-posteriori truths. They will be like:

> The poison most beloved of mystery writers = arsenic.

On the other hand, if the connection between functional role and mental state is a reference-fixing one, the identities between mental states and brain states will be necessary a-posteriori ones like (according to most philosophers and I won't be arguing the toss here):

> Water = H_2O.

However, although this is a big difference, as far as the topic of this volume goes, I suspect it is a small difference. On both views, enough information about brain state and functional role will determine without remainder what subjects do (modulo indeterminacy, a matter we will address shortly), and will do so in a way that allows one to predict what subjects will do. Supporters of the two views will disagree about which parts of the body of information are crucial and why, and how the functional part relates to the neurological part, but they will agree on the point that if you tell me enough about John Doe's internal functional roles and brain processes, you tell me what determines what he will do, and you do so in a way that allows me to predict what he will do.

Now I need to mention a complication that, I am glad to report, causes less trouble than one might have feared.

6. The distinction between action and movement

There is an important distinction between behaviour in the sense of what a person does, and behaviour in the sense of the movements of the person's body (raw behaviour, as it is sometimes called). My reaching for a glass is something I do. In doing it, my arm will move through space in some way that brings my hand near to the glass. The action and the movement are not the same. Indeed, the same action can involve different movements: there are many ways of hailing a taxi, and the same movement can be part of different (intentional) actions: for example, turning a light on in Australia and turning a light off in America require the same downward movement. This invites the question, were we talking about actions or movements in the previous few paragraphs?

However, for the two kinds of physicalism we were discussing, it doesn't matter – that's why the complication doesn't cause trouble. According to physicalists of either of these two kinds, enough detail about the causal history of the movements of a person's body delivers automatically the causal history of the person's actions. The reason is that, for these physicalists, each and every action is nothing over and above some movement of the body with the right causal history in terms of mental states. This means that, for them, the causal history of a body's movements and the causal history of the actions of a person with that body are not two different topics. Although classifying by actions differs from classifying by movements, what a person does is determined without remainder by, and is fully predictable given enough information about, physical antecedents alone – or rather this is true given determinism. It is time to discuss the implications of indeterminism in physics.

7. The implications of indeterminacy

It is very likely that determinism is false[3] but it turns out that this makes less difference than one might have expected for the implications of physicalism for the possibility of free action.

There are two cases to consider. One is that, as far as mental states, actions and bodily movements go, determinism might as well be true. This

[3] Although experts in quantum theory tell me that interpretations of quantum theory in which determinism is true are undergoing something of a revival.

is because, at the macroscopic level of actions, of those brain states which are mental states, and of bodily movements, in the vast majority of cases things proceed as if determinism were true. The indeterministic effects at the sub-microscopic level wash out as we aggregate. Obviously, in this case the truth of indeterminism is by the way in the debate over free will.

The other, more likely case (as I understand matters but I defer to experts in physics) is that there *is* significant indeterminacy at the level of actions, of those brain states which are mental states, and of bodily movements. In this case, what persons do at a time is not determined by, and is not fully predictable from, how things are beforehand with their brains and surroundings even if physicalism is true. However, although what they do today isn't fully determined by or predictable from how things were beforehand in terms of their brains and functional states, the *chances* of their doing this, that or the next thing are fully determined by, and predictable from, how their brains are at a given earlier time. What they in fact do today then has an irreducibly chancy connection with those earlier chances. Although physicalists must deny that the way my brain plus surroundings are, here and now, fully determine everything I do in the future, they should affirm that the way my brain plus surroundings are, here and now, fully determine the chances for every possible future thing I do, and that the passage from a chance of my doing X to my subsequently doing X is irreducibly chancy.

A natural thought is that in this second case, the case where the indeterminacy at the sub-microscopic level doesn't wash out as we aggregate, the truth of indeterminism has significant implications for whether or not physicalism poses a threat to the existence of free action. This, however, turns out to be a mistake. The reason goes back to the overall nature of the debate over free will and determinism.

In the centuries-long debate over whether the hypothesis that what we do at any given time is fully determined by, and predictable from, how things were before that time, does or does not imply that we never act freely, the warring parties divide into two camps. One camp, that of the compatibilists, says that determinism is compatible with acting freely on occasion. Their key argument is that determinism is consistent with an action's being under an agent's control, in the sense that the agent would have done otherwise if they had chosen to do otherwise. For, they note, determinism is compatible with the causal path from the past to an action having, as a *crucial* component, the agent's decision. What is more, they observe, determinism is consistent with the action's *reflecting* the agent's desires and character. For determinism is consistent with an action's being caused by an agent's desires and character. And, according to compatibilists, something like the foregoing two facts, or some suitable tweaking of them, is all that needs to be the case for an action to count as a free action.[4] Now, all this remains true when we factor in chances. As far as the point about control is concerned, the only change is that our actions will be under our control in the sense that had we chosen otherwise, the chances would have been different, and, moreover, the chances will normally be very close to one or zero as the time of action approaches. As far as the point about the connection between, on the one hand, desire and character, and action, on the other, goes, it remains true that there is a distinction between actions that reflect one's desires and character and those that don't. Making the connection between action and character and desires to some extent chancy doesn't alter this.

Those in the other camp – those who hold that determinism is incompatible with acting freely – insist that all this talk of control, of the causal path to action going via an agent's decision, of an action's possibly reflecting the agent's character and desires, is focusing on the wrong question, is indeed a piece of misdirection, on the part of compatibilists. The challenge that determinism poses to the existence of free actions concerns the causal *antecedents* of a person's decisions, desires and character, and the like. Those in the incompatibilist camp allow that determinism is consistent with some actions being such that had the agent decided to do otherwise, they would have done otherwise. So what, they say. The key question concerns the causal origins of the decisions agents in fact make. They are unlikely to find it comforting to be told that these origins include a chancy element. Likewise, incompatibilists grant that some actions reflect the desires and characters of those who perform them but insist that the key question, when

---

[4] See, e.g., Ayer (1954). Recent defences of compatibilism – often influenced in one way or another by Frankfurt (1971) – have introduced significant modifications but they don't affect our point here and below. The modifications respond to the point that human beings can stand in complex relationships to their desires, their decisions and their characters. For example, human beings can have desires and characters they would prefer not to have, or ones they judge to be, in some sense, irrational, or ones that they, in some sense, disown. Many smokers arguably fall into one or more of these categories. This suggests that compatibilists should specify what it is to act freely in terms of acting in accord with some suitably *restricted* set of desires, decisions and characters – for example, the ones that agents themselves want to have. But none of this affects the point that whether or not there is an element of chance is by the way in the debate.

our topic is the existence of free actions, is the causal origins of the desires and characters. They worry that our desires and our character come from factors outside our control – our genetic make-up and the conditions into which we were born, for example. But discovering that the connection between our genes and our desires and character is in part chancy isn't the discovery that we have control over our genes or the economic circumstances of our parents. In sum, incompatibilists urge that the challenge posed by determinism to the existence of free action lies in what determinism says about the causal antecedents of an agent's decisions, character and desires, and the like; the crucial point being that those causal antecedents are outside an agent's control. (For a version of this line of thought, see Strawson, this volume.) Bringing chance into the picture doesn't somehow make these past factors into ones that are under the agent's control.

I am saying nothing about which party, the compatibilist or the incompatibilist, wins the debate. Our point is that it is hard to see how inserting chances into the picture could affect who wins, one way or the other.

8. Anomalous physicalism

We have been discussing the implications of a version of physicalism that puts identities between mental states and brain states centre stage. There is, however, a very different version of physicalism, a version that holds it is a mistake in principle to identify the mental and the physical, in the sense of identifying mental and physical *kinds*. At first blush, this kind of physicalism might seem good news for those who worry about the implications of physicalism for freedom. I'll be saying that the good news is not that good.

This kind of physicalism starts from the notion of a physical property, relation, law and entity, where they are of a kind with those to be found in physics, chemistry and biology. But this isn't a definition of the physical. It is an *indication* of the kinds of properties etc. intended. The official definition of the physical is simply that it is whatever we need to give a full account of the *non*-sentient. The thesis of physicalism is then the claim that our world is nothing over and above a huge aggregation of elements that are themselves non-sentient. The sentient emerges from this aggregation in the same way that a house emerges when you put the various bits, none of which is itself a house, together in the right way. We know that aggregation creates new patterns. That's the fun in playing with Lego. The idea, in the case of the mind, is that, when we suitably aggregate the non-sentient, some of the new patterns emerging from the aggregation will be mental properties. Indeed, the idea is that exactly this happened when each and every one of us was conceived. Conception triggered a complex process of aggregation of bits that are, in themselves, non-sentient to deliver over time something that is sentient.

(This definition makes a potentially controversial assumption about the foundations of quantum mechanics. The assumption is that talk of the observer in quantum mechanics is a case of bad labelling. Anyone who reads about the two-slit experiment will be struck by the thought that there is something deeply puzzling going on and should hope that smart physicists will sort the mess out for us. One possible sorting out will give the observer *as such* an important, fundamental role. Would this mean that sentience is a fundamental ingredient of our world? That seems unlikely. Despite all the uncertainty about what to say about the two-slit experiment, it seems that the role of the observer is to be a measuring device, something big enough to carry information about what is going on. Sentience as such isn't part of the story.)

So far we have not said anything that necessarily goes against an identity version of physicalism – the identity version can agree that we are nothing over and above aggregations of the physical, in the sense of physical we have been discussing. The clash comes when we add the distinctive thesis of anomalous physicalism – or, rather, the distinctive thesis of a 'vanilla' version of anomalous physicalism. Our interest is in the core idea that underpins a range of views about the relation between the physical and the psychological that agree in repudiating dualism while insisting that the kind of reductionist picture that comes from functionalism or the type-type identity theory is mistaken. (For versions of this core idea, see Davidson 1980a, 1980b, and the references in Davies 2000, and especially his discussion of 'upwards explanatory gaps'; for some dissent from the core idea, see Jackson 2010.) The distinctive thesis is that there are no patterned dependencies running from the physical to the mental. What exactly does this come to?

Physicalism is committed to the supervenience of the mental on the physical. If the mental is nothing over and above a complex aggregation of elements that are purely physical, then duplication in the physical implies duplication in the mental. This means there exists a raft of true conditionals of the form

If X is in P_i, then X is in M_j

which take us from a full enough specification of X's physical nature to the mental state that supervenes on that physical nature.

However, this is consistent with the absence of a patterned dependence of the mental on the physical. Supervenience *per se* only requires that, for each physical antecedent, there is exactly one mental state. It does not require that, for each mental state, the physical antecedents of the conditionals with that mental state as consequent have any degree of unity. The claim of anomalous physicalism is that if we group the conditionals together by sameness of mental consequent, we won't find any pattern uniting the physical antecedent that is sufficient to allow us to identify mental types with physical types.

How might one argue for anomalous physicalism? We cannot argue in the standard way from the problems for epiphenomenalism about mental properties. We cannot, for example, offer an argument of the following form:

> Pain is causally efficacious with respect to behaviour (denial of epiphenomenalism)
>
> All causally efficacious properties with respect to behaviour are physical properties (the causal closure of the physical)
>
> Therefore, pain is a physical property (by transitivity)

Anomalous physicalism precisely denies that there are any true identities between mental and physical properties. We have to appeal instead to a difference principle. It says that the mental makes a difference. Had there been no mental properties, things would have been very different behaviourally. This claim is consistent with denying that mental properties are causally efficacious. Take the property of being fragile. Plausibly, though contestably, being fragile is not a causally efficacious property. What makes a dropped glass break is not its being fragile but is instead the categorical basis of its fragility – the thinness and internal structure of the glass, as it might be. Maybe mental properties are like that; indeed many say exactly this. All the same, being fragile may make a difference in the following sense: it may well be true that had the glass not been fragile, it would not have broken. Those who hold that being fragile isn't causally efficacious don't go around telling people that it is fine to knock fragile glasses over.

How do we use the difference principle to argue for physicalism? By arguing that we know enough about what the world is like and how it works, to be confident that the only way things might have been different in the relevant behavioural ways is if they were different physically. Thus, if being different mentally makes for behavioural differences, this can only be because any and every given mental difference corresponds precisely to some physical difference. But then physical sameness implies mental sameness. What is the best explanation of this striking fact? The thesis that the mental emerges from the aggregation of physical.

9. Does anomalous physicalism change things that much?

A natural first thought is that the whole issue looks very different if one embraces anomalous physicalism. No longer does one have the spectre of patterns in the physical imposing themselves on the mental, and especially on those aspects of the mental especially concerned with intentional action. In some sense, we get a genuine autonomy of the mental and the intentional but in a way that avoids implausible metaphysical 'additions' of the kind that make so many uncomfortable with dualism. And it is worth recalling that this general idea has a substantial history that predates modern discussions of physicalism (see, e.g., Ryle 1954).

Here's the problem; the bad news. We noted earlier the distinction between movement and action. For identity versions of physicalism the distinction isn't of great moment as far as the prediction of intentional action goes. This is because, as we observed, full information about physical antecedents of movements delivers full information about actions according to identity versions of physicalism. This, however, is not the case if anomalous physicalism is true. Although the physical determines the mental, if anomalous physicalism is true, it will often be the case that we cannot, as a matter of principle, infer the mental from the physical. We will have, as we might say it, determination without *predictable* determination of the mental by the physical. In this sense, the mental will be autonomous. In this sense, we have some 'wriggle' room that might suggest a way of making sense of free action. This is because an agent's intentional actions, what the agent *does*, won't be fully predictable going by the physical alone, and the same goes for the chances of the agent's doing this or that. However, and this is where the bad news comes in, it remains true that full information about physical antecedents delivers full information about *movements* and full information about the subsequent distribution of physical objects. But if you worry that the predictable determination of what you do, or the chances of what you do, at any given time, by a purely physical past would rob you of freedom of action, it would seem ad hoc not to worry that the predictable determination of the movements of your body, or of the location of objects in your vicinity, or the chances of same, by a purely physical past robs you of freedom of movement or of the freedom to rearrange how things are around you. If your movements aren't free, what freedom of action remains? It would be strange to celebrate how free one's actions are, while conceding that you aren't free to cause your eyelid to close, or to change the angle your arm is at by some given number of degrees, or to change the location of a chess piece on a chess board. A gift of freedom of intentional action that comes without freedom of movement or freedom to change things around one would not be much of a present.

10. The upshot

What does physicalism imply about the determination of action? We have seen that there is no standard answer. It all depends on the correct version of physicalism, and on the implications of indeterminism in physics. However, perhaps surprisingly, we have also seen that physicalism has no particular implications for the existence of free action, or so I have argued.

References

Armstrong, D.M. (1968) *A Materialist Theory of the Mind*, London: Routledge & Kegan Paul.

Ayer, A.J. (1954) Freedom and necessity. In *Philosophical Essays*, New York: St Martin's Press, pp. 3–20.

Davidson, D. (1980a) Mental events. In *Essays on Actions and Events*, Oxford: Oxford University Press, pp. 207–25.

Davidson, D. (1980b) The material mind. In *Essays on Actions and Events*, Oxford: Oxford University Press, pp. 245–59.

Davies, M. (2000) Interaction without reduction: the relationship between personal and sub-personal levels of description, *Mind and Society*, 1(2): 87–105.

Frankfurt, H. (1971) Freedom of the will and the concept of a person, *Journal of Philosophy*, 66: 829–39.

Jackson, F. (2010) The autonomy of mind, *Philosophical Issues*, 20 (Philosophy of Mind), pp. 170–84.

Ryle, G. (1954) *Dilemmas*, Cambridge: Cambridge University Press.

Williamson, T. (2009) Replies to critics. In P. Greenough and D. Pritchard (eds) *Williamson on Knowledge*, Oxford: Oxford University Press, pp. 279–384.

4
Dualism and the determination of action

RICHARD SWINBURNE[1]

I argue in this chapter that it is most unlikely that neuroscientists will ever be able to predict human actions resulting from difficult moral decisions with any high degree of probable success. That result leaves open the possibility that humans sometimes decide which actions to perform, without their decisions being predetermined by prior causes. I need to begin with two assumptions, which provide a different framework within which to work out how far human actions are predictable from that of Frank Jackson (see the previous chapter), and which lead to a different kind of conclusion. I have space here only to provide brief justifications of these assumptions; for fuller justifications I must refer readers to other writings of mine.

1. Brain events and mental events interact

My first assumption (not held by Frank Jackson) is that there are goings-on (unchanging states or changes of states) of two non-overlapping kinds, ones which are public (i.e. equally accessible to all), and ones to which their subject has privileged access. I shall call these goings-on 'events'; the former I shall call 'physical events' and the latter 'mental events'. Physical events include brain events; anyone can discover as well as can anyone else what is going on in my brain. But my having a headache is an event to which I have privileged access, and so it is a mental event. Someone else can learn

[1] Many thanks to Daniel Robinson and to two anonymous reviewers for very useful comments on an earlier draft of this paper.

about my headache from my behaviour and from studying my brain (in the sense that studying these can show them that it's quite probable that I have a headache); but I also could learn about my headache in these ways, and yet I have a further means of learning about it by actually experiencing the headache. Some mental events do however have physical events as a constituent part. My seeing my desk is a mental event, but its occurrence entails that there is a desk present (a physical event). I define a pure mental event as one which does not entail the occurrence of a physical event – that is, it is not part of what is meant by the claim that that event occurs that some physical event occurs, although it is compatible with the claim that that event occurs that it is caused by a physical event. My headache is a pure mental event. (In another terminology pure mental events are 'narrow content' mental events.) All my subsequent references to mental events are to be understood as references to pure mental events.[2]

Among such mental events, some are such that necessarily if and when we have them we are (at least to some degree) conscious (aware) that we are having them. This group of mental events includes not merely sensations such as pains (often called 'qualia'), but also (occurrent) thoughts (which may be entertained without being believed). If I am not in any way aware that the thought 'today is Saturday' is now 'crossing my mind', it isn't crossing my mind. But my definition includes as mental events also events which still exist while the subject is not conscious of them, but of which the subject may become conscious from time to time. In this group are desires (inclinations to do some action to which the subject may or may not yield) and beliefs. I can desire to get home in time for lunch, or to write a great book – while not thinking about this or doing anything to achieve my desire. I have at any time lots of beliefs – about history or geography for example – of which I am not in any way conscious at that time. My concern in this chapter with intentions (that is, purposes) is solely with the intentions in our present intentional actions, for example my intention in walking along a certain road being to walk to the railway station, not with

---

[2] For a full justification of the (to many of us) obvious point that there are these two kinds of 'going-on', see Swinburne (1997: Part I). This first assumption is the assumption of property (and so event) dualism, which is the moderate kind of dualism. The more radical kind of dualism, substance dualism, which I also advocate but do not assume in this chapter, is discussed in Howard Robinson's chapter in this volume. For the superiority of the way of making the mental/physical distinction in terms of a subject's privileged access, over other ways, see Swinburne (2007: 142–4 and nn. 3 and 4).

intentions to be executed later; and I shall understand by an 'intentional action' an action which one knows one is doing and means to do. In that sense of 'intention' each of us is always to some degree conscious (that is, aware) of our intentions. When we are to some degree conscious of any of our beliefs, intentions, and desires then (like sensations and thoughts), they are conscious events. And we are to some extent conscious of many such mental events all the time. For example all perceptions (and we perceive things all the time we are awake) involve not merely (or primarily) having sensations but consciously acquiring beliefs.

My second assumption is that – despite the recent work of neuroscientists pursuing the programme pioneered by Benjamin Libet and discussed (in their chapters in this volume) by Patrick Haggard and Tim Bayne – not merely are many conscious events caused by brain events, but conscious events often cause brain events and other conscious events. (Frank Jackson agrees that 'mental events' cause brain events, but he does not understand 'mental' events in the same way as I do.) My reason in brief for assuming that Libet-type experiments do not show that intentions do not in general cause the brain events which cause the bodily movements involved in intentional actions is that the evidence adduced by neuroscientists includes (as evidence necessary for establishing their conclusion) evidence about when subjects form various intentions (e.g. to move a hand). This evidence comes from what subjects tell the scientists about when they believe that they formed some intention. But scientists would only be justified in believing what the subjects tell them if they believed that subjects say what they do because they intend to tell the truth about their beliefs; that is, the scientists must believe that a subject's intention (together with a belief) causes him to say what he does. If they thought that the words coming out of a subject's mouth were caused only by a sequence of brain events themselves not caused by any intention to tell the truth, they would have no justification for believing what a subject tells them. From that it follows that, while some experimental results might show that sometimes intentions do not cause brain events, reaching that conclusion requires the assumption that often (e.g. when subjects tell scientists what they believe) intentions do cause brain events (and so bodily movements). Most of our beliefs about the world (everything most of us ever learnt about history or geography or science, including all the experimental evidence adduced by scientists and the scientific theories based on it) are derived from what other people tell us, and we believe what they say because we believe that an intention to tell the truth about what they believe causes them to say what they do. On the assumption that we are right to believe many of the things which we are told, we must assume that our informants' intentions, together with the beliefs implied by what they say, cause brain events which cause bodily movements.[3]

Intentions to perform basic actions (ones which we do, not by doing any other action), such as moving a hand or uttering a sentence, can no doubt cause effects without needing to be combined with many, if any, other conscious events (such as conscious beliefs) in order to do so. But most intentions are intentions to perform non-basic actions; a non-basic action is an action which an agent does by doing some other action – for example when I walk to the railway station by walking along a certain road. An intention to perform a non-basic action (e.g. to walk to the railway station) needs to be combined with a belief about which basic actions (e.g. of walking along a certain road) will lead to that intention being fulfilled.

As well as assuming that many conscious events cause brain events, I also assume that conscious events sometimes cause other conscious events in the kind of rational way codified by theorists of practical and theoretical reasoning, as when I perform a series of calculations leading to a belief about the result of a complicated sum, or when my belief that the coin has landed heads 99 out of 100 times causes my belief that it will land heads next time. Not merely does this seem fairly evidently so, but without this assumption we would have no justification for believing any of our inferences. If I am to be justified in believing some conclusion which I have reached by considering some argument, I must believe that I am moved (that is, caused) to believe it by reflecting on the earlier stages of the argument. Without this assumption we would have to think of ourselves as incapable of reaching conclusions (e.g. that some scientific theory is true) on the basis of evidence.

So when we consciously form an intention (i.e. make a decision), we are often influenced by conscious beliefs (including value beliefs) and conscious desires. Value beliefs (as I shall understand this notion) are beliefs about the overall objective (including moral) goodness of doing different actions. In so far as I believe an action good to do I have a reason and so some inclination to form the intention to do it. While other beliefs need to be combined with some desire or prior intention in order to motivate us to act, value beliefs as such motivate us.[4] If I believe that it is obligatory to

---

[3] For full argument in justification of this claim see Swinburne (2011).

[4] For argument in support of this point, see Swinburne (1998: Additional note 3).

keep a promise, I will have some inclination to keep the promise; I couldn't think of it as obligatory if I did not. And the better I believe some action to be, the greater as such is my inclination to do it. But a value inclination may be weak, and I may yield instead to some other inclination.

A desire to do some action, as I shall understand this notion, is an inclination to form an intention to do the action when the subject believes that possible, independently of any inclination caused by a belief that it is objectively good to do it. Desires in this sense are a matter of what we 'feel like' doing. Beliefs and desires are, at a given time, involuntary states; I cannot change them at will. I believe that today is Saturday, that I am now in Oxford, that Aquinas lived in the thirteenth century, and so on and so on. I cannot suddenly decide to believe that today is Monday, that I am now in Italy, or that Aquinas lived in the eighteenth century. (Although I cannot change a belief at will, I can set about investigating a topic (e.g. when Aquinas lived), which might (but might not) lead to a change of belief.) Likewise we find ourselves desiring sleep or food, fame or fortune; we cannot (normally) change these desires suddenly, but we can decide not to do the action which they incline us to do, and we can take steps which may lead to a change of desire in the future.

If I have no desire to do one available action rather than another (e.g. to give money to this charity rather than that charity), I form the intention to do what I believe is the best action to do, that is the one which, it seems to me, I have most reason to do. If I have no belief about the relative value of certain actions (e.g. to lunch at this restaurant rather than that one), I form the intention to do that available action which I most desire to do, that is the one which I find myself most inclined to do. But when I have equally strong desires to do each of two incompatible actions (and no stronger desire to do a different incompatible action) and no relevant value belief, or a belief that two incompatible actions would be equally good (and no other incompatible action would be better) and no relevant desire, and – above all – when what I most desire to do is incompatible with what I believe to be the best action to do, which intention I will form cannot be determined solely by the relative strengths of my desires or my beliefs about which action would be best to do. If I am caused to decide (as opposed to deciding without being caused to decide) to form an intention in such circumstances, the final outcome must be determined by brain events in a non-rational way.

I have argued that beliefs and desires are caused, and I shall assume that all other mental events (conscious or not conscious) with the possible exception of intentions are also caused. Clearly some desires and sensations are caused directly by brain events without help from any other mental events; desires to drink or sleep, and sensations of pain or noise are surely in this category. But most of our desires, and – I suggest – all our beliefs and occurrent thoughts couldn't be had without coming in mutually sustaining packages of other beliefs and desires; or be conscious without being sustained by other conscious beliefs and desires. I could not desire to be prime minister without this desire being sustained by many beliefs about what prime ministers do, as well no doubt as some brain events causing me to desire to be famous or powerful. And I couldn't even come consciously to believe (through perceiving it) that there is a book on the table in front of me without having many other conscious beliefs, such as a belief that books are written by authors for people to read, and a belief that tables have flat surfaces, and so on.

In order to simplify the discussion, I shall assume that even if some mental (including conscious) events need other mental (including conscious) events to sustain them, a subject's total conscious state (all his conscious events) at a given time (with the possible exception of his intentions) is caused ultimately (sometimes via earlier conscious or other mental events) by his total brain state; and also that every type of total conscious state (with the same possible exception) is correlated with some type (or disjunction of types) of overall brain state. (So when a conscious event causes another conscious event, either the former or the latter causes a brain event correlated with the latter.) Most total conscious states will be large ones, full of beliefs and sensations (consider the many perceptual beliefs involved in coming to see a scene), and often some occurrent thoughts, desires or intentions. The brain state which is correlated with a conscious state will also normally be a large state; recent neuroscience suggests that it consists in a 'temporal synchrony between the firing of neurons located even in widely separated regions of the brain', between which there are 'reciprocal long-distance connections', a synchrony which attains a 'sufficient degree and duration of self-sustained activity'.[5] Different overall conscious states are correlated with different variants of this pattern of activity. So if we are to make predictions of future conscious events and brain events, we would need a theory of which aspects of a total brain state (which types of individual brain events) cause or are caused by which

---

[5] Gray (2004: 173 and 175). The 'global workspace' model has been confirmed by recent work of Raphael Gaillard and others; see Robinson (2009).

aspects of a total mental (including conscious) state. Then we could predict that any new total brain state which contained a certain type of brain event would cause a certain type of conscious event, such as a certain type of intention; or conclude instead that some intentions occur uncaused.

2. Obstacles to assembling data for a mind-brain theory

To construct such a mind-brain theory we would need a lot of data in the form of a very long list of particular (what philosophers call 'token') conscious (and other mental) events occurring simultaneously with token brain events. To get information about which conscious events are occurring, we depend on the reports of subjects about their own conscious events. There are however two major obstacles which make it difficult or impossible to get full information from subjects.

The first obstacle concerns the 'propositional' mental events, occurrent thoughts, desires, beliefs, and intentions. I call them 'propositional events'[6] because they involve an attitude to a proposition (which forms the content of the attitude). A belief is a belief that such-and-such a proposition is true; a desire is a desire that such-and-such a proposition be true, and so on. The problem is that while the content of most of these events can be described in a public language, its words are often understood in slightly different senses by different speakers. One person's thought which he describes as the occurrent thought that scientists are 'narrow-minded', or the belief that there is a 'table' in the next room, has a slightly different content from another person's thought or belief, described in the same way. What one person thinks of as 'narrow-minded' another person doesn't, and some of us count any surfaces with legs as 'tables' whereas others discriminate between desks, sideboards, and tables.[7] This obstacle can be overcome, by questioning subjects about exactly what they mean by certain words. But it has the consequence that, since beliefs etc. are the beliefs they are in virtue of the way their owners think of them, far fewer people have exactly the

---

[6] They are sometimes called 'intentional' events, but I avoid this label since it leads to confusion between intentions and the wider class of 'intentional events'.

[7] Though it hardly needs such support, this point is borne out by recent experiments showing that presenting some image to a group of subjects produced in all subjects similar patterns of activity in different regions, but slightly different patterns for each subject. See Shinkareva *et al.* (2008).

same particular beliefs, desires, etc. as anyone else than one might initially suppose – which makes the kind of experimental repetition which scientists require to establish their theories much harder to obtain. And it seems most unlikely that any two humans understand all their words in exactly the same way, and so have exactly the same concepts as each other.

There is however a much larger obstacle to understanding what people tell us about their sensations. This is that we can understand what they say only on the assumption that the sensations of anyone else are the same as we would ourselves have in the same circumstances – and that is often a highly dubious assumption. This obstacle applies to all experiences of colour, sound, taste and smell (the 'secondary qualities'). We can recognize when someone makes the same discriminations as we do between the public properties of colour etc., but we cannot check whether they make the discriminations on the basis of the same sensations as we do. While it might seem counter-intuitive to suppose that green things look to one group of people just like red things look to another group of people, while red things look to the first group just like green things look to the second group, other possibilities seem less counter-intuitive. Maybe green things look a little redder to some people than they do to others,[8] or coloured things look fainter to some people than to others, when neither of these differences affect their abilities to make the same discriminations.

We could rule out such possibilities on grounds of simplicity (that it is simpler and so more probable to suppose that these things do not happen), if it were the case that which neurons have to fire in which sequence at which rate in order to produce a sensation which subjects call 'green' (or whatever) were exactly the same in all humans. But in view of the differences between the brains of different humans, that seems very improbable. It's much more likely that sometimes for two different people different neurons produce a sensation which they both call 'green'. The different reactions which people often have to the same input from the senses supports the hypothesis that the sensations caused thereby are sometimes different in different people. Some people like the taste of curry, others don't. There are two possible hypotheses to explain this: curry tastes the same to everyone but some people like and some people don't like this taste, or curry tastes differently to different people. It would seem highly

---

[8] Even if two groups of subjects typically agree in the percentage of 'redness' shown by greenish colour samples, that won't show that the 'pure green' or 'pure red' samples look the same to both groups, and so that a '10% red' sample looks the same to both groups.

arbitrary to suppose that the first explanation is correct – let alone suppose that a similar explanation applies to all different reactions to tastes.

I need however to make a qualification to all this, that while we may be unable to understand the natures of the individual sensations of others, their sensations may exhibit patterns which are the same as some publicly exemplifiable patterns (of primary qualities such as shape). Thus a mental image of a square has the same shape as a public square. The lines which make up the image may have peculiarities of colour which the subject cannot convey, but he can convey the shape. I shall return to this point later.

3. Obstacles to forming a predictive theory from the data

So, bearing in mind these limits to the kinds of mental data we can have, what are the prospects for forming a theory supported by evidence which will not merely explain and so predict how brain events cause sensations, beliefs, and desires, but how these (together with brain events) cause our subsequent intentions?

What makes a scientific theory such as a theory of mechanics able to explain a diverse set of mechanical phenomena is that the laws of mechanics all deal with the same sort of thing – material objects – and concern only a few of their properties – their mass, shape, size, and position – which differ from each other in measurable ways (one has twice as much mass as another, or is three times as long as another). Because the values of these measurable properties are affected only by the values of a few other such properties, we can have a few general laws which relate two or more such measured properties in all objects by a mathematical formula. We do not merely have to say that, when an inelastic object of 100g mass and 10m/sec velocity collides with an inelastic object of 200g mass and 5m/sec velocity, such and such results, with unconnected formulas for the results of collisions of innumerable inelastic objects of different masses and velocities. We can have a general formula, a law saying that for every pair of inelastic material objects in collision the quantity of the sum of the mass of the first multiplied by its velocity plus the mass of the second multiplied by its velocity is always conserved. But that can hold only if mass and velocity can be measured on scales – for example, of grams and metres per second. And we can extend mechanics to a general physics including a few more measurable quantities (charge, spin, colour charge, etc.) which interact with mechanical quantities, to construct a theory which makes testable predictions.

A mind-brain theory however would need to deal with things of very different kinds. Brain events differ from each other in the chemical elements involved in them (which in turn differ from each other in measurable ways) and in the velocity and direction of the transmission of electric charge. But mental events do not have any of these properties. The propositional events (beliefs, desires, etc.) are what they are, and have the influence they do in virtue of their propositional content, often expressible in language but a language which – I noted earlier – has a content and rules differing slightly for each person. (And note that while the meaning of a public sentence is a matter of how the words of the language are used, the content of a propositional event such as a thought is intrinsic to it; it has the content it does, however the subject or others may use words on other occasions.) Propositional events have relations of deductive logic to each other; and some of those deductive relations determine the identity of the propositional event. My belief that all men are mortal wouldn't be that belief if I also believed that Socrates was an immortal man; and my thought that '2=1+1, and 3=2+1, and 4=3+1' wouldn't be the thought normally expressed by those equations if I denied that it followed from them that '2+2=4'. And so generally, much of the content of the mental life cannot be described except in terms of the content of propositional events; and that cannot be done except by some language (slightly different for each person) with semantic and syntactic features somewhat analogous to those of a public language. The rules of a language which relate the concepts of that language to each other cannot be captured by a few 'laws of language' because the deductive relations between sentences and so the propositions which they express are so complicated that it needs all the rules contained in a dictionary and grammar of the language to express them. These rules are independent rules and do not follow from a few more general rules. Consider how few of the words which occur in a dictionary can be defined adequately by other words in the dictionary, and so the same must hold for the concepts which they express; and consider in how many different ways describable by the grammar of the language words can be put together so as to form sentences with different kinds of meaning, and so the same must hold for the propositions which they express.

So any mind-brain theory which sought to explain how prior brain events cause the beliefs, desires, etc. which they do would consist of laws relating brain events with numerically measurable values of transmission of electric charge in various circuits, to conscious (and non-conscious) beliefs, desires, intentions, etc. with a content individuated by sentences of a language (varying slightly for each person), and also sensations. The contents of mental events do not differ from each other in any numerically measurable way, nor do they have any intrinsic order (except in the respect that some contain others – e.g. the belief about the book contains a belief about its uses). Those concepts which are not designated by words fully defined by other words – and that is most concepts – are not functions of each other. And they can be combined in innumerable different ways which are not functions of each other, to form the propositions which are the contents of thoughts, intentions, etc. So it looks as if the best we could hope for is an enormously long list of separate laws (differing slightly for each person) relating brain events and mental events without these laws being derivable from a few more general laws.[9]

Could we not at least have an 'atomic' theory which would relate particular brain events involving only a few neurons to particular aspects of a conscious state – particular beliefs, occurrent thoughts, etc., the content of which was describable by a single sentence (of a given subject's language), in such a way that we could at least predict that a belief with exactly the same content would be formed when the same few neurons fired again in the same sequence at the same rate (if ever that happened)? The 'language of thought' hypothesis[10] (LOT), which takes seriously the analogy of the brain to a computer which manipulates symbols, seems to involve some version of an atomic theory. It claims that there are rules relating brain events and beliefs of these kinds, albeit a very large and complicated set of them. It holds that different concepts and different logical relations which they can have to each other are correlated with different features in the brain. For example, it holds that there are features of the brain which are correlated with the concepts of 'man', 'mortal', and 'Socrates', and that there is a relation R which these features can sometimes

---

[9] Donald Davidson is well known for arguing that 'there are no strict psychophysical laws' (1980: 222). This thesis is stronger than mine, but his reasons for it are similar to mine. However he uses this thesis in defence of his theory of 'anomalous monism', that all events are physical, some events are also mental, and so physical-mental causal interaction (which we both recognize) is law-like causal interaction of two physical events. But, contrary to Davidson, I am assuming (for reasons stated briefly at the beginning of this chapter) that there are events of two distinct types, physical and mental; and so I reject Davidson's resulting theory.

[10] Originally put forward by Fodor (1979).

have to each other. When someone believes that Socrates is mortal, R holds in their brain between the 'Socrates'-feature, and the 'mortal'-feature; when someone believes that Socrates is a man, R holds between the 'Socrates'-feature, and the 'man'-feature; and when someone believes that all men are mortal, R holds between the 'man'-feature and the 'mortal-feature'. (The holding of this relation might perhaps consist in the features being connected by some regular pattern of signals between them.) The main argument given for LOT is that unless our brain worked like this, the operation of the brain couldn't explain how we reason from 'all men are mortal' and 'Socrates is a man' to 'Socrates is mortal', since our reasoning depends on our ability to recognize the relevant concepts as separate concepts connected in a certain particular way. Beliefs, and so presumably other propositional events, the theory claims, correspond to 'sentences in the head'.

I argued earlier however that no belief can be held without being sustained by certain other beliefs – for logical reasons; which other beliefs a given belief is thought of as entailing determines in part which belief the latter belief is. Now consider two beliefs, whose content is expressed in English by 'this is square' and 'this has four sides'; someone couldn't hold the first belief without holding the second. So these two beliefs cannot always be correlated with different brain events, since in that case a neuroscientist could eliminate the brain event correlated to the latter belief without eliminating the brain event correlated to the former belief. On the other hand these two beliefs cannot always be correlated with the same brain event since someone can have the belief 'this has four sides' without having the belief 'this is square'. It follows that propositional events are correlated with more than one (type of) brain event. That leads to the view that propositional events only occur as part of a total mental state, including many other mental events, and it is this whole mental state which is correlated with a whole brain state without there being correlations between separate parts of the mental and brain states. This view is that of connectionism,[11] the rival theory to LOT. Mind-brain relations are holistic. Only if connectionists hold, as they often do, that mental events are identical with (or supervene logically on) individual brain events, is it an objection to connectionism that brain events do not have a structure corresponding to that of a human language. But given my initial assumption

[11] For a selection of papers on both sides of the language-of-thought/connectionism debate see Parts II and III of Lycan and Prinz (2008).

that mental events are events distinct from brain events, mental events can have a sentential structure without brain events having this. So, given connectionism, a mind-brain theory could at best only predict the occurrence of some conscious event in the context of a large mental state (consisting of many beliefs, desires, etc., some of them conscious) and of a large brain state (events involving vast numbers of neurons).

We have seen that we must suppose that mental events often cause other mental events in a rational way. As I illustrated earlier, the laws of rational thought include the criteria of valid deductive inference, and – since the validity of an inference between sentences depends on which propositions are expressed by which sentences, and that depends on the meanings and arrangements of the words the sentences contain – these can only be stated fully by lists as long as those of the dictionary and grammar of a human language. These laws also include the criteria of cogent inductive inference (that is, criteria of inductive probability, of which propositions make which other propositions probable). They also include the criteria for forming value beliefs. But each human person has slightly different criteria of these kinds. Further, of course, humans do not always follow their own criteria of rational thought, and so we would need laws stating when and how brain events disturb rational processes. These latter laws would vary with the overall mental and brain states of the subject, and the mental states which disturb rationality (such as beliefs) would often need to be described in terms of the concepts with which that subject operates (e.g. some particular fixation preventing someone reasoning rationally about a particular subject matter).

Further, insofar as mental events cause other mental events in a rational way, the influence of mental events depends on their strength; and (apart from occurrent thoughts) they all have different strengths. One person's sensation of the taste of curry is stronger than another person's. One person's belief that humans are causing global warming is stronger than another person's (that is, the first person believes this proposition to be more probable than does the second person). Yet while subjects can sometimes put sensations in order of strength in virtue of their subjective experience, what they cannot do – despite 150 years of work on 'psychophysics' – is to ascribe to them numerical degrees of strength in any objective way.[12] (How do I answer the doctor who asks 'Is this pain more

[12] See Laming (2004): 'Most people have no idea what "half as loud" means. In conclusion, there is no way to measure sensation that is distinct from measurement of the physical

than twice as severe as that pain?') And, despite 80 years of work on 'subjective probability', the same applies to beliefs and other propositional events.[13] Such differences affect behaviour in a rational way. Someone is more likely (despite counter-influences) to stop eating curry, the stronger is the taste of curry which she dislikes. Someone is more likely (despite counter-influences) to choose to travel by bus rather than by car because of a belief that humans are causing global warming and a belief that it is good to prevent this, the stronger are those beliefs. In order to measure the influence of sensations, beliefs, etc. on intentions (in a situation where there are many conflicting influences), we need a measure of their absolute strength which can play its role in an equation connecting these; and subjects cannot provide that from introspection. Neuroscience might discover that greater frequency of certain kinds of brain event causes the beliefs caused by those brain events to be stronger. But for prediction of their effects we'd need to know how much stronger were the resultant beliefs. So we'd need a theory by means of which to calculate this, which gave results compatible with subjects' reports about the relative strengths of their beliefs. But the brain circuits, rates of firing, etc., which sustain beliefs in different subjects are so different from each other that it is difficult to see how there could be a general formula connecting some feature of brain events with the strength of the mental events which they sustain. So the most we could get is a long list of the kinds of brain activity which increase or decrease the strength of which kinds of mental events.

So the part of a mind-brain theory which predicts human intentions and so human actions would consist of an enormous number of particular

stimulus.' The latter sentence is a bit pessimistic; it might be possible to measure it from the (in some sense) strength of the brain event which caused it. But there is no reason to suppose that a noise which was by such a measure half as loud as a second noise would seem that way to its hearers.

[13] There is a long tradition beginning with the work of F.P. Ramsey, of attempting to measure someone's degree of belief in a proposition (the 'subjective probability' which they ascribe to it) by the lowest odds at which they believe that they would be prepared to bet that it was true. If someone is, they believe, prepared to bet £N that q is true at odds of 3-1 (i.e. they would win £3N if q turned out true, but lose their £N if q turned out false) but not at any lower odds (e.g. 2-1), that – it was claimed – showed that they ascribe to q a probability of 1/4 (because then in their view what they would win multiplied by the probability of their winning would equal what they would lose multiplied by the probability of their losing). But that method of assessing subjective probability will give different answers varying with the amount to be bet – someone might be willing to bet £10 at 3-1 but £100 only at odds of more than 4-1; and people have all sorts of reasons for betting or not betting other than to win money.

laws relating brain events to subsequent sensations, thoughts, beliefs, and desires (some of them conscious), and these (together with other brain events) to subsequent intentions, having this kind of shape:

> Brain events $(B_1, B_2. . .B_j)$ + sensations $(M_1. . .M_e)$ + Thoughts $(M_f. . .M_i)$ + Beliefs (including value beliefs) $(M_j. . .M_k)$ + Desires $(M_k. . .M_l)$ → Intention (M_n) + Beliefs (about how to execute the intention) $(M_p. . .M_q)$ + Brain events → bodily movements.

The B's describe events in individual neurons, and each law would involve large numbers of these; the M's describe particular mental events with a content describable by a short sentence, and with a strength. The strength of an intention measures how hard the agent will try to do the intended action.

In cases where mental events alone determine the resulting intention, we can no doubt often predict that intention in virtue of the general principles outlined at the beginning of the chapter – e.g. where there is a strongest desire and no relevant value beliefs, or a strongest value belief and no relevant contrary desire, and the agent believes that he is able to do the action, the agent will form the intention to act on the strongest desire or value belief (even if we cannot predict whether the intention will be strong enough to be executed). But where brain events interact with mental events to form desires and beliefs and thereby to determine subsequent intentions, the previous argument has the consequence that – if determined – the outcome would be determined for each person by one of an enormous number of different laws relating total brain states to total mental states, including large total conscious states. So we could not work out what a person will do on one occasion when she had one set of brain events, beliefs and desires, on the basis of what she (or someone else) did on a previous occasion when she had a different set differing only in respect of one relevant belief. For there would be no general rule about the effect of just that one change of belief on different belief and desire sets; the effect of the change would be different according to what was the earlier set, and what were the brain states correlated with it. But no human being ever has the same total brain state and mental state at any two times or the same total brain state and mental state as any other human does at any time, and – I suggest – no human being considering a difficult moral decision ever has the same conscious state, let alone the same brain state in the respects which give rise to consciousness and determine its transitions, as at another time or as any other human ever. For making a difficult moral decision involves taking into account many different conflicting beliefs and desires. The believed circumstances of each such decision will be different, and (consciously or unconsciously) an agent will be much influenced by her previous moral reflections and decisions.

Consider someone deciding how to vote at a national election. She will have beliefs about the moral worth of the different policies of each party, and the probability of each party executing its policies; she will desire to vote for this candidate and against that candidate for various reasons (liking or disliking them for different reasons); she will desire to vote in the same way as (or in a different way from) her parents, and so on. But each voter will have slightly different beliefs and desires of these kinds. So because each voter's total conscious state would never have occurred previously, there could not be any evidence supporting a component law of the mind-brain theory to predict what would happen this time. A very similar conscious state might have occurred previously (in the same or a different voter), supporting a detailed law about the effects of that similar conscious state. But that suggested law (weakly supported by one piece of evidence) would (because of the slight difference in the conscious state) only predict what would happen this time with a degree of probability surely less than a half. Add to this the point that the part of the overall brain state which determines the strength of the different events constituting the conscious state, and how rationally subjects will react to them, will almost certainly be different on the different occasions. Add to all this the points made in Section 2 about the difficulty involved in getting some of the evidence required to support any mind-brain theory, and I conclude that it is most unlikely that a prediction about which difficult moral decision someone would make, and so which resulting action they would do, could ever be supported by enough evidence to make it probably true. Human brains and human mental life are just too complex for humans to understand completely.

That conclusion has a crucial consequence that those brain events which cause the movements which constitute human actions will never be totally predictable. But even if it should turn out that the behaviour of other physical systems is totally predictable, it should not be too surprising that the brains of humans (and perhaps higher animals) are different, since the brain is unlike any other physical system in that it causes innumerable non-physical events.

4. What neuroscience can discover

The limits to the ability of neuroscience to predict arise from the enormously large number of detailed laws which would have to govern any interaction of many different kinds of mental (including conscious) events and brain events. But neuroscience may be able to discover, and has begun to discover, mind-brain laws which do not involve such complicated interactions. Thus it has begun to discover which particular brain events are necessary and sufficient for the occurrence of those non-propositional events which do not involve the inaccessible aspects of sensations, but only the patterns of sensations. A mental image has the same sort of properties of shape and size as the properties of public objects such as brain events. So neuroscience is on the way to discovering a law-like formula by which it can predict from a subject's brain events both the images caused by the public objects at which she is looking and the images which she is intentionally causing.[14] But that formula will not tell us what the subject regards their image as an image of – e.g. as an image of a television set or of a shiny box. Which beliefs subjects acquire about what they are seeing is clearly going to vary with their prior beliefs about the way objects of different kinds look, e.g. that something of such and such a shape is a television set. But if the neuroscientist discovers these prior beliefs in some other way (e.g. from what subjects tell him, or by analogy with his own beliefs), then he should be able to predict from a subject's brain events not merely the shape of the image which she is having, but the subject's belief about what she believes that she is seeing.

Similar considerations apply to the other senses. Which words a subject hears depends on the pattern of sensed sounds rather than their intrinsic qualities; and patterns of sensed sounds have the same describable shape as patterns of public sounds. So it should be possible to construct a formula describing how the brain events caused by certain patterns of public sounds cause patterns of sensed sounds. Given people's linguistic beliefs (their beliefs about what words mean) discoverable in some other way, it should then be possible to predict from their brain events what they understand to be the content of what is being said to them. So scientists should be able to arrange for sentences to be 'heard' by the deaf whose auditory nerves

---

[14] K.N. Kay *et al.* (2008) devised a decoding method which made it possible to identify, 'from a large net of completely novel natural images, which specific image was seen by an observer'.

no longer function, by means of electrodes in their brain causing the appropriate brain events.

Desires to do basic actions can occur in the absence of a large set of beliefs. Hence neuroscience can discover the brain events which are the immediate causes of desires to form intentions to do basic actions, such as to drink. It can also discover the brain events which are the immediate effects of intentions to perform basic actions (e.g. move a hand, or utter a certain word), these being intentions which can be had independently of any other beliefs. That will enable it to detect what 'locked in' people are trying to do, and so set up some apparatus which will enable them to succeed.[15] But in order to predict which non-basic action a subject has the intention of performing, a neuroscientist would need to know the subject's beliefs about which basic actions would bring about the performance of the non-basic action. Hence we need to know subjects' linguistic beliefs in order to know which proposition as opposed to which words (defined by the sounds which constitute its utterance) subjects are trying to utter.

Neuroscience may be able to make various kinds of statistical predictions, to the effect that a change in the pattern of certain kinds of brain events will probably lead to an increase or decrease in the strength of certain kinds of desire or belief and so to the probability of certain intentions. Thus it may be able to discover how certain brain events affect the relative strengths of very general kinds of desire (e.g. for fame or power). Desires influence but, when the subject also has competing particular desires and value beliefs, do not determine a subject's intentions and so behaviour. And which intention a general desire will tend to cause will depend on the subject's beliefs (e.g. about how fame can be obtained). So again in the absence of a formula for calculating beliefs of any complexity from brain events, and in the absence of a formula for calculating intentions from competing beliefs and desires (and brain events), all we can hope for is statistical predictions to the effect that the more or less of some physical quantity that brain events have, the greater or less the desire to do so-and-so, and so – probably – the greater the proportion of subjects who will do so-and-so. Hence drugs or mirror neurons may indeed promote or diminish altruistic desires,[16] or strengthen or weaken a desire to commit suicide.

---

[15] See the work (described in Kellis *et al.* 2010) being done to detect the brain events caused by intentions to utter certain sounds, which will enable computers to translate these into speech, when people are 'locked in' and unable to communicate by speech.

[16] Paul J. Zak *et al.* (2009) found that increasing testosterone in men makes them less generous in the game situations created by psychologists.

But such increases or decreases of desires only yield statistics; they don't tell you who will do what, since we all have different rival desires of different strengths and different value beliefs of different strengths.

It follows however, finally, that neuroscience should be able to predict what individual humans will do in order to execute certain general instructions which have as a consequence that their behaviour must depend on only one simple desire of a kind caused directly by a brain event. For example, in the Libet (2004: ch. 4) experiments discussed earlier in this volume subjects were told to move their hand at any time within a short period when they decided to do so; and since they would not have had any value beliefs about when to do so, they must have decided to do so when they 'felt like' it, i.e. desired to do it. Such a desire is like an itch and so presumably has a direct cause in a brain event. So in this case neuroscience may be able to correlate prior brain events with the movements which they cause, via the desire to cause them. If however subjects disobeyed the instructions, and didn't move their hand within the period – either because they didn't feel the requisite desire or because they had rival desires (e.g. to be a nuisance) or value beliefs (e.g. that it was immoral to take part in the experiment), their actions would not count in assessing the experiment. So 100% success in predicting hand movements under these experimental conditions is by no means impossible. But once again that tells us nothing about how people will behave in situations of conflicting desires and value beliefs.

So despite the possibility (and in some cases the actuality) of all these advances in neuroscience, the main point of this chapter remains, that for the prediction of individual behaviour in circumstances where there are different variables, both brain events and mental events of different and competing kinds and strengths affecting the outcome, neuroscience would need a general formula well supported by evidence to enable it to relate the strengths of these kinds of events to each other; and that cannot be had.

5. Intentions are probably undetermined

Yet, even if it is unpredictable which intention we will form in such circumstances and how strong it will prove, what reason do we have for supposing that that intention (with its particular strength) is not caused (in a way too complicated to predict) by brain events? After all, I have acknowledged, our intentions are often caused – when they are caused by a strongest desire and we have no contrary value belief, a strongest value belief and we have no contrary desire, and when our desires and value beliefs are in opposition to each other. My answer is that it is just under the circumstances where desires or value beliefs are of equal strength or in opposition to each other, that we are conscious of deciding between competing alternatives. We believe that it is then up to us what to do, and we make a decision. Otherwise, as we realize, we just move along habitual paths. It is a basic principle of rationality that things are probably the way they seem to be (in the sense that we are inclined to believe that they are) in the absence of counter-evidence. All science begins from experience, and experience is experience of the way things seem to be (in the physical or mental world). When we make a decision, it seems that we choose and are not caused to choose as we do; in other cases it does not seem that we are making a choice which is up to us there and then. So, in the absence of counter-evidence (in the form of a causal theory of our behaviour in such circumstances, rendered probably true by much evidence), when we make a decision, we are probably doing so without being caused to do so.

References

Davidson, D. (1980) Mental events. In *Essays on Actions and Events*, Oxford: Oxford University Press.

Fodor, J.A. (1979) *The Language of Thought*, Cambridge, MA: Harvard University Press.

Gray, J. (2004) *Consciousness: Creeping up on the Hard Problem*, Oxford: Oxford University Press.

Kay, K.N. *et al.* (2008) Identifying natural images from human brain activity, *Nature*, 452: 352–5.

Kellis, S. *et al.* (2010) Decoding spoken words using local field potentials recorded from the cortical surface, *Journal of Neural Engineering*, 7: 1–10.

Laming, D.R.J. (2004) Psychophysics. In Richard L. Gregory (ed.) *The Oxford Companion to the Mind*, Second edition, Oxford: Oxford University Press, pp. 771–3.

Libet, B. (2004) *Mind Time*, Cambridge, MA: Harvard University Press.

Lycan, W.G. and Prinz J.J (eds) (2008) *Mind and Cognition: An Anthology*, Third edition, Oxford: Blackwell.

Robinson, R. (2009) Exploring the 'global workspace' of consciousness, PLoS Biology 7(3): doi10.1371/journal.pbio.1000066.

Shinkareva, S.V. *et al.* (2008) Using fMRI brain activation to identify cognitive states associated with perception of tools and dwellings, PLoS ONE 3(1): e1394.doi10.1371/journal.pone.0001394.

Swinburne, R. (1997) *The Evolution of the Soul*, Revised edition, Oxford: Oxford University Press.

Swinburne, R. (1998) *Providence and the Problem of Evil*, Oxford: Oxford University Press.

Swinburne, R. (2007) From mental/physical identity to substance dualism. In P. van Inwagen and D. Zimmerman (eds) *Persons: Human and Divine*, Oxford: Oxford University Press.

Swinburne, R. (2011) Could anyone justifiably believe epiphenomenalism?, *Journal of Consciousness Studies*, 18(3/4): 196–216.

Zak, Paul J. *et al.* (2009) Testosterone administration decreases generosity in the ultimatum game, PLoS ONE, 4(12): e8330.doi:10.1371/journal.pone.0008330.

5
On determinacy or its absence in the brain

HARALD ATMANSPACHER AND STEFAN ROTTER

1. Introduction

A number of philosophers nourish the hope that studies of brain behaviour will resolve the question whether the brain is a completely deterministic machine, or a generator of stochastic states with indeterminate connections.[1] Although brain science is mainly concerned with how particular phenomena are described best, philosophical issues often refer to the real, ontic nature of these phenomena.

In this paper we analyse the different ways to describe brain behaviour with the goal to provide a basis for an informed discussion of the nature of decisions and actions that humans perform in their lives. As will become clear, this is a difficult task, and we are far from solving it with currently available knowledge. But it is worthwhile to locate some major stumbling blocks and address approaches intended to remove or circumvent them.

Much of the material that we are going to present has been extracted from a previous publication (Atmanspacher and Rotter 2008) which we recommend for readers who desire to see more details. Here we try to reduce formal and technical details to what we think is necessary for proper understanding. Moreover, we restrict ourselves to brain states and their dynamics exclusively and exclude any relation to the mental. An earlier review with similar focus has been published by Glimcher (2005).

---

[1] We use the terms 'stochasticity' and 'stochastic behaviour' as synonymous with 'randomness' and 'random behaviour'. The notion of randomness seems to be favoured in the Russian tradition, whereas stochasticity is more frequently used in the Anglo-American terminology.

The article is organized in a straightforward fashion. In Section 2 we outline a number of concepts exhibiting how many subtle details and distinctions lie behind the broad notions of determinacy and stochasticity. These details are necessary for a sound and fertile discussion, in Section 3, of particular aspects relevant for the characterization of brain states and their dynamics.

Our basic notion is that the descriptions of brain behaviour currently provided by neuroscience depend on the level and context of the descriptions. There is no clear-cut evidence for ultimately determinate or ultimately stochastic brain behaviour. As a consequence, we see no solid neurobiological basis to argue either in favour of or against any fundamental determination or openness of human decisions and actions.

2. Between determinacy and stochasticity

In this section some elementary concepts are introduced that are crucial for an informed discussion of determinacy and stochasticity, but are often used with vague associations, or badly defined specifications, or both. The terms *determinateness* and *determinism* are used to distinguish between the determinacy of *states* of a system (or classes of such states) and its *dynamics* (or classes of its dynamics), respectively.[2] By way of characterizing these terms, closely related notions such as error and variation, causation and prediction are addressed which serve important purposes particularly in the description of complex systems. Then the concepts of stochasticity, randomness, chance, and the role that probability plays in their context are outlined, and techniques of how to represent stochastic processes deterministically and vice versa are described.

An important further issue is the difference between *individual* and *statistical* descriptions of a system. It is tightly related to one of the most fundamental questions concerning the status of scientific theories: the distinction between epistemological and ontological statements (Scheibe 1974; Atmanspacher and Primas 2003). Epistemology comprises all kinds of issues related to the knowledge (or ignorance) of information gathering and using systems. By contrast, ontology refers to the nature and behaviour

---

[2] The terminology is adopted from Jammer's discussion of Heisenberg's usage of these terms (Jammer 1974).

of systems as they are (or may be assumed), independent of any empirical access.

2.1 State determinateness

States of a system to which *epistemic descriptions* refer are called *epistemic states*. The mathematical representation of such states encodes empirically obtainable knowledge about them. If the knowledge about states and their associated properties (also called observables) is expressed by probabilities in the sense of relative frequencies for a statistical ensemble of independently repeated experiments, we speak of a *statistical description* and of *statistical states*. Insofar as epistemic descriptions refer to a state concept encoding knowledge, they depend on contexts of observation and measurement. Therefore, descriptions of properties or observables associated with an epistemic state are *contextual*, i.e. context-relative.

States of a system to which *ontic descriptions* refer are called *ontic states*. They are designed as (idealized) exhaustive representations of the mode of being of a system, i.e. an ontic state is 'just the way it is', without any reference to what any observer does or does not know about it. The properties (or observables) of the system are understood as *intrinsic*. As *individual states*, ontic states are the referents of *individual descriptions*. The ontic state of a system is represented by a point x in its state space Ω. The associated intrinsic observables define the coordinates of the state space. The state of a system is *determinate* if it is point-wise represented and, as a consequence, the observables are sharply defined.

An epistemic (statistical) state is characterized by a probability measure μ over Ω. Such a representation of epistemic states (and their associated observables) generally requires a partition of Ω into subsets A. It refers to our knowledge as to whether an ontic (individual) state x is more likely to be in some subset A rather than in others. This limit on the informational content of an epistemic state μ can be due to two basically distinct reasons. First, it can be due to 'objective' influences such as uncontrollable perturbations originating from the system's environment or intricate interdependencies among its internal constituents. Second, it can be due to 'technical' reasons such as the imprecision of measurements or the fact that any decimal expansion of real numbers has to be truncated somewhere for computational purposes.

In the second case, the epistemic state encodes the reduced amount of knowledge about an individual ontic state that results from explicitly epistemic procedures. We use the notion of *error* to address this situation. Error characterizes the scatter of measured or rounded values of an observable with respect to an assumed 'true' value in an ontic state. Variation, on the other hand, refers to a distribution of ontic states, each of which is characterized by a 'true' value of one of its associated observables. This is the situation in the first case above.[3] In complex systems, epistemic states and the corresponding distributions generally include both error and variation.

In quantum systems, the situation is even more complicated since even observables associated with ontic states are usually not dispersion-free, meaning that ontic quantum states are generally indeterminate. This is related to the non-commutativity of observables in quantum theory, giving rise to 'blurred observables' (Schrödinger 1935) that prevents a point-wise representation of an ontic quantum state. This idea of an 'objective vagueness' has been picked up under terms like 'Verschmiertheit', 'properties lacking sharp values', 'ontic blurring', 'objective fuzziness', 'unsharp properties'. This must not be confused with fluctuations or statistical spreads, amounting to variations due to supposed valuations by point functions.

2.2 Dynamic determinism

The temporal evolution of an ontic state $x \in \Omega$ as a function of time $t \in \mathbb{R}$ is a trajectory $t \mapsto x(t)$; the ontic state $x(t)$ determines the intrinsic observables that a system has at time t exhaustively. The temporal evolution of an epistemic state μ corresponds to the evolution of a subset $A \subset \Omega$ over time. The concept of an individual trajectory of an individual, ontic state is lost within a purely epistemic description.

The evolution is time-reversal invariant, if for every solution $f(t)$ of the equations of motion also the function $f(-t)$ is a solution. In this case, the evolution of the system is called both forward and backward *deterministic*. In such a case, there is no preferred direction of time. Fundamental physical laws (e.g. in Newton's mechanics, Maxwell's electrodynamics, relativity theory) are time-reversal invariant in this sense. (Note that fundamental physical laws are also time-translation invariant, such that no instance in time is distinguished and, thus, there is no nowness in physics.)

---

[3] From the perspective of statistical modelling, the two situations are known as fixed-effect modelling (with errors) versus random-effect modelling (with variation). Within a stochastic approach, the latter case is sometimes characterized as doubly stochastic.

Phenomenological physical theories such as thermodynamics contain a distinguished direction of time where the time-reversal symmetry of fundamental laws is broken. This leads to an evolution that is either forward or backward deterministic. The breakdown of both time-translation invariance and time-reversal invariance is necessary to distinguish first a particular instant in time t (nowness) and then, relative to this instant, two temporal domains called past and future. In this way, a *tensed time* can be generated which differs fundamentally from the *tenseless parameter time* of physics.

For a discussion of notions such as *causation* and *prediction*, concepts that are more specific than the concept of determinism, tenseless time alone is insufficient. Among several varieties of causation, this applies particularly to *efficient causation* in the sense of cause-effect relations that are conceived on the basis of a well-defined ordering relation between earlier and later states. This ordering is also significant for the notion of prediction (and retrodiction).

Specifically intricate types of dynamics arise in classical systems that are today known as deterministic chaos. The behaviour of chaotic systems in this sense is governed by deterministic equations, yet it is not predictable to arbitrarily high accuracy because, roughly speaking, it depends sensitively on the accuracy with which initial conditions are known. This is sometimes popularly referred to as the so-called 'butterfly effect', but a much better example is the mechanical three-body problem, e.g. in the motion of celestial bodies. Quite innocently looking equations of motion, simply integrable for two gravitationally coupled bodies, become exceedingly difficult to solve if only one body is added.

In the theory of complex systems, chaotic behaviour is characterized by an intrinsic instability with respect to changes of initial conditions. This instability can be quantified by so-called Lyapunov exponents or, more compactly, by the Kolomogorov-Sinai entropy of the system. Positive values of these quantities indicate that small variations of initial conditions become exponentially amplified as a function of time. They define a finite 'predictability horizon', giving rise to the notion of 'weak causation' for chaotic systems. For instance, the orbits of many smaller bodies of the solar system (asteroids, moons), subjected to the combined gravitational perturbations of the major planets, are chaotic and unstable on million-year timescales.

But even in the absence of chaos, indeterministic processes may occur at particular instances of time, whose simplest prototypes are bifurcations.

When a stable, non-changing state of a system becomes unstable, it typically relaxes into one among several neighbouring stable states at a so-called critical bifurcation parameter. Various kinds of bifurcations are known today – their common characteristic is that small fluctuations at the critical point play a significant role for the behaviour of the system beyond that point. In a general sense, such behaviour is generic for all kinds of decision processes where a selection (choice) among two or more alternatives is at stake.

The time evolution of the state of a quantum system is strictly deterministic if it is described by the Schrödinger equation. Issues about determinacy arise if individual quantum states are considered. For instance, measurement processes on quantum systems in superposition states are not describable by the deterministic Schrödinger equation – in this sense they are indeterministic. Although the epistemic state of the system after measurement is determinate, the ontic superposition state prior to measurement is indeterminate in the sense that the observables characterizing it are not dispersion-free.[4]

On the other hand, the applicability of the Schrödinger time evolution is, strictly speaking, limited to the case of isolated systems. For the dynamics of open quantum systems, stochastic influences by the environment have to be taken into account. This, however, is usually done by classical noise rather than genuine quantum stochasticity.

2.3 Probability and stochasticity

The concept of determinism cannot only be contrasted with causation and prediction. An important arena with its own tradition is the tension between deterministic and stochastic behaviour. Roughly speaking, stochastic behaviour is what is left if one tries to describe the world in deterministic terms.

In the theory of stochastic processes, the indeterministic behaviour of a system is described in terms of stochastic variables $x(t)$, parametrized by time $t \in \mathbb{R}$, whose distribution $\mu(x, t)$ represents an epistemic state. The description of a system in terms of individual trajectories $x(t)$ corresponds

---

[4] Note that here and elsewhere in this article we refer to states and observables of quantum systems rather than quantum events. The reason is that there is no well-defined and consistent notion of an event in quantum theory. This makes it notoriously difficult to relate philosophical discussions focusing on events to quantum theory and its results.

to a point dynamics of an ontic state, whereas a description in terms of the evolution of the associated measure $\mu(x, t)$ corresponds to an ensemble dynamics of an epistemic state.

The limiting cases of infinite and vanishing predictability indicated in the preceding section correspond to special types of stochastic transition matrices. For instance, singular stochastic processes are completely deterministic and allow a perfect prediction of the future from the past. The general case of limited predictability is covered by the notion of a regular stochastic process. Particular complications arise where sequences of states or observables are not stationary, or where limit theorems of probability theory (central limit theorem, law of large numbers) do not apply. This is typically the case in complex systems.

The dichotomy of ontic and epistemic descriptions is also prominent in the theory of stochastic differential equations. For instance, Langevin-type equations generally treat stochastic contributions $\xi(t)$ in addition to a deterministic flow in terms of fluctuations around the trajectory of a point x in phase space. Such a picture clearly reflects an ontic approach. On the other hand, the evolution of epistemic states μ, e.g. in terms of drift and diffusion, is typically described by Fokker-Planck-type equations.

2.4 Deterministic embedding of stochastic processes

It is not surprising that deterministic processes such as fixed points or periodic cycles can be considered as special cases of more general formulations in terms of stochastic processes. This is easily understandable if all entries in a stochastic transition matrix are either 1 or 0, thus representing deterministic transitions among states. What comes somewhat as a surprise is the converse, namely that stochastic processes can be understood in terms of deterministic processes. This has been accomplished by means of a mathematical theory of so-called natural extensions or dilations of stochastic processes (see Gustafson 1997 for an overview).[5]

The significance of this important result can be illustrated in the following way. A system that is originally described stochastically, e.g. due to uncontrollable interactions with its environment, can be extended into its environment until all relevant interactions are integrated in the

---

[5] Note: this theory proves that deterministic embeddings of stochastic processes exist; it does not give an explicit and unique prescription of how to construct them.

behaviour of the system itself. This leads to an increasing number of degrees of freedom, enabling an integration of all previously stochastic behaviour into an overall deterministic dynamics. The price to be paid is a phase space of considerably higher dimensions, which has to cover all degrees of freedom that were treated as noise in the stochastic formulation.

If deterministic and stochastic dynamics can be rigorously transformed into each other, it follows that corresponding *(epistemic) descriptions* cannot be taken as indicators of determinacy and stochasticity as *(ontic) features of nature.* The dilation theorems mentioned above show that the gap between the two is often a matter of convenience or convention. If a stochastic description of a system seems to be appropriate, this can be due to deliberately regarding interactions with its environment as noise, or due to lacking knowledge about underlying observables and corresponding laws of nature. Conversely, certain idealizations that ignore fluctuations or perturbations of a system can lead to a deterministic description.

3. Brain behaviour: deterministic or stochastic?

The brain is a complex network comprising a very large number of inter-connected nerve cells (neurons), and an equally large number of supporting cells (glia cells). Neurons send signals to each other in the form of electrical impulses (action potentials), using specialized extensions of their cell body (axons and dendrites). The axon of the sending neuron forms specialized junctions (synapses) with the dendrites of the receiving neurons. At the synapse, a presynaptic action potential triggers the release of neuro-transmitters (e.g. glutamate) into the synaptic cleft. The transmitter is then bound to specific receptors in the postsynaptic neuron. This, in turn, induces a brief electrical signal in the postsynaptic neuron, which can be either excitatory or inhibitory, depending on the transmitter-receptor pair used for signal transduction. The balanced action of thousands of excitatory and inhibitory inputs to every neuron determines its activation and, in view of the high degree of recurrency found in most parts of the brain, also its contribution to the overall brain dynamics.

Different aspects of neurodynamics are salient for brain activity at different levels of resolution, as for example ion channels, neurons, or networks of neurons. Therefore, the corresponding conceptual frameworks and mathematical models present a picture of neurodynamics that is quite non-uniform. Sweeping from microscopic over mesoscopic to macroscopic components of the brain, deterministic, chaotic or stochastic descriptions can each be adequate. The proper choice for one of them is usually motivated by arguments concerning which features are important and which aspects have little or no relevance at each particular level.

This entails that it is often premature – and always risky – to draw general conclusions from neurodynamics for the organism by taking only a single level of description into account. In fact, an interpretation derived from models at one level may be contrary to the conclusions obtained from another level. Different scenarios at the level of stochastic subcellular biophysics, at the level of quasi-deterministic isolated neurons and at the level of large chaotic networks yield different conclusions.

A number of issues that were, and still are, intensely discussed at the level of individual neurons (often referred to as the micro-level of brain activity) will be outlined in Section 3.1. Although ion channels, which are essential for the functioning of neurons, are best described stochastically, the reliability and precision of individual neurons suggests a deterministic picture. Section 3.2 addresses the complex dynamics of recurrent cortical assemblies or networks (of several thousand recurrently coupled neurons), characterized by apparently irregular activity but presumably reflecting deterministically chaotic dynamics. This so-called mesoscopic level of brain activity is the level at which neural correlates of mental representations are assumed.[6]

3.1 Ion channels and individual neurons

Stochastic dynamics are prevalent in brain activity observed at a microscopic level. Consequently, probabilistic methods play a prominent role for modelling phenomena in this realm. In particular, the essential molecular components of electrical activity in nerve cells, the so-called ion channels, open and close stochastically and are effectively described by Markov processes (Sakmann and Neher 1995). Such models assume a probability (rate) associated with each possible transition, and modulatory

---

[6] The neural correlate of a mental representation can be characterized by the fact that the connectivities, or couplings, among those neurons form an assembly confined with respect to its environment, to which connectivities are weaker than within the assembly. A simple mechanism for the formation of such assemblies was proposed by Hebb (1949) and has been considerably refined by the concept of spike-timing dependent plasticity (Gerstner *et al.* 1996; Markram *et al.* 1997) in recent years.

parameters (like the membrane potential at the spike initiation zone, or the concentration of neurotransmitters in the synaptic cleft) may influence these transition probabilities. Almost-determinate (almost-deterministic) behaviour is reflected by state (transition) probabilities close to 0 or 1, respectively.

The function of intact neurons relies on large populations of ion channels rather than few individual ones. The compound ion currents resulting in larger membrane patches and whole neurons have greatly reduced fluctuations as compared to single-channel currents. This averaging or smoothing of signals associated with (lower-level) ion channel populations is due to the independent, uncorrelated operation of its constituents. As a consequence, the classical (higher-level) description of single neurons and their dynamics including action-potential generation is both determinate and deterministic, first described by Hodgkin and Huxley (1952) in terms of non-linear deterministic differential equations. Not only was this formal description found very adequate for isolated nerve cells, subsequent physiological experiments also demonstrated that real neurons can be extremely reliable and precise. In fact, it was demonstrated that neurons effectively operate (under certain conditions) as deterministic spike encoders (see Bryant and Segundo 1976; Mainen and Sejnowski 1995).

Both notions, reliability and precision, address epistemic issues of predictability horizons and the form of distributions. They avoid ontic connotations or commitments which need to be clarified if one speaks about determinacy and stochasticity. Reliability and precision are useful to express the relevance of particular phenomena on a microscopic level (e.g. stochastic ion channel openings) for the dynamics on a more macro-scopic level (e.g. the operation of nerve cells), and the degree of control that can be exerted either by the system itself or by an experimenter. It cannot be denied, though, that chance phenomena characterize the operation of some subcellular building blocks of neurons in an essential way. These aspects, however, do not seem to play an important role for the functioning of neurons, and even less so for the dynamics of the large networks that they constitute.

This is not surprising to the degree to which an extension of a system into its environment often allows to simplify the description of its dyna-mics, e.g. by averaging procedures. As indicated in Section 2.4, this can lead to a deterministic description if enough stochastic interactions are integrated, i.e. if the system is considered globally enough. In this way, the stochastic dynamics of ion channels underlying the behaviour of individual neurons 'averages out' so as to yield a highly precise and reliable deterministic description of neurons.

Any distribution characterizing a state can in principle be interpreted as due to genuine chance (ontically) or due to ignorance (epistemically). This ignorance, or missing information, can in turn be deliberate, e.g. in order to disregard details that are inessential in a particular context, or it can be caused by uncontrollable perturbations. At the ion channel level, where quantum effects must be expected to occur, an ontic interpretation in terms of indeterminate states is possible or likely.

However, the fact that the stochastic dynamics of ion channels typically yields highly reliable and precise neuronal behaviour suggests that any potentially genuine micro-stochasticity is inefficacious at the neuronal level (and even more so at the network level). Therefore, statistical neuronal states are assumed to be of epistemic nature and genuine indeterministic contributions to the dynamics of neurons seem to be of low relevance. After all, the representation of the neurodynamics in a neuronal state space amounts to a fairly well-defined trajectory of quasi-ontic states.

3.2 Neuronal assemblies and recurrent networks

The activity of cortical neurons *in vivo*, i.e. embedded in networks, is generally characterized by highly irregular spike trains (Softky and Koch 1993). The standard deviation of inter-spike intervals is typically close to the mean interval, a property which is also shared by Poisson processes (cf. Tolhurst *et al.* 1981). This reminiscence of a random phenomenon raises the question which components of single-neuron activity can be considered as a signal, and which aspects must be classified as random noise (Shadlen and Newsome 1998).

Detailed models of input integration in neurons conceive the membrane potential process as a continuous-valued stochastic process like a one-dimensional random walk (Gerstein and Mandelbrot 1964; Tuckwell 1988). Irregular spike trains then correspond to strongly fluctuating membrane potentials observed in intracellular recordings *in vivo* (Azouz and Gray 1999). Comparison with *in vitro* recordings (most afferents cut off) strongly suggests that irregular spike trains *in vivo* are actually caused by complex, but deterministic synaptic interactions among the neurons in the network, and that they are not a result of noise processes extrinsic or intrinsic to the network.

Different from the many kinds of stochastic or probabilistic modelling of brain dynamics, observations of simple deterministic dynamics, e.g. periodic processes such as in EEG signals during epileptic seizures, are exceedingly rare. However, the irregularly looking complex dynamics of recurrent networks of neurons comprising both excitatory and inhibitory cells have been identified as features of deterministic chaos. The idea to model neurodynamics in terms of deterministic chaos goes back to early work of Freeman (1979); see also Amit (1989). It has been of increasing influence over the decades (see, e.g., van Vreeswijk and Sompolinsky 1996; Jahnke *et al.* 2009), and today there is general agreement that deterministically chaotic processes play an important role to understand the dynamics of the brain at various levels.

The dynamics of a chaotic system takes place on a subspace of the state space that is called an attractor. However, this holds only *cum grano salis*, namely if the transient motion from initial conditions on to the attractor can be disregarded. If such transient motion is not negligible, or if there are many attractors between which the system spends considerable amounts of time, the situation becomes quite complicated. For instance, Kaneko (1990) observed *long-time (super-) transients* in studies of complex coupled map lattices, and Tsuda (2001) proposed the scenario of *chaotic itinerancy* with many coexisting quasi-stable 'attractor ruins' for neurodynamics. Recent work of Zumdieck *et al.* (2004) demonstrates that long chaotic transients may abound in complex networks.

In such involved situations, observed irregularities of time series are due to a combination of deterministic chaos and random fluctuations which are hard to disentangle. It is, therefore, often practical to look for an overall stochastic description, for instance in terms of diffusion equations, stochastic point processes, or otherwise. Corresponding analyses (based on a Fokker-Planck-type diffusion approximation) of single-neuron dynamics in the high-input regime showed qualitatively different types of mean field dynamics of recurrent networks, depending on the strength of external inputs and on the relative strength of recurrent inhibition in the network (Brunel and Hakim 1999).

However, this does not imply that the nature of the process is fundamentally stochastic. Mixtures of deterministic and stochastic processes can obviously be described by stochastic evolution operators, although there may be underlying features of hidden determinism. As outlined in Section 2.4 above, even the opposite is possible: systems that are described deterministically can have underlying features of hidden stochasticity as well. But no matter whether the described features are hidden or apparent, they remain epistemic and do not justify conclusive ontic implications.

3.3 Quantum stochasticity in neural activity: a pertinent example

Since stochastic descriptions can generally be transformed into deterministic descriptions, it is an important issue whether or not there is a *genuinely* stochastic neuronal dynamics for which a (hidden) deterministic interpretation can be excluded. To all our present knowledge this is the case for stochastic quantum processes, for which the Schrödinger dynamics breaks down so that they are to be conceived as ontologically indeterministic.

Although the neurobiological literature does not refer to quantum phenomena a lot, it would be premature to assume that indeterministic quantum processes do not occur on relevant scales in the brain. A fairly concrete and detailed suggestion of how stochastic quantum processes could play a role is due to Beck and Eccles (1992), later refined by Beck (2001). This proposal refers to particular mechanisms of information transfer at the synaptic cleft.

The information flow between neurons in chemical synapses is initiated by the release of transmitters in the presynaptic terminal. This process, called exocytosis, is triggered by an arriving nerve impulse with some small probability. Thermodynamics or quantum mechanics can be invoked to describe the trigger mechanism in a statistical way. Examining the corresponding energy regimes shows that quantum processes are distinguishable from thermal processes for energies higher than 0.01 eV at room temperature (Beck and Eccles 1992). Assuming a typical length scale for biological microsites of the order of several nanometers, an effective mass below 10 electron masses is sufficient to ensure that quantum processes prevail over thermal processes.

The detailed trigger mechanism proposed by Beck and Eccles (1992) is based on the quantum concept of quasi-particles, reflecting the particle aspect of a collective mode. Skipping the details of the picture, the proposed trigger mechanism refers to tunnelling processes of two-state quasi-particles, resulting in indeterministic quantum state collapses. It yields a probability of exocytosis in the range between 0 and 0.7, in agreement with empirical observations.

However, processes at single synapses can hardly be correlated with 'active' decision processes, whose neural correlates are widely assumed to be coherent assemblies of neurons. The same problem appears if, according to Eccles's original intention, it is suggested that quantum processes in the brain provide an entry point for mental states to act causally on neural states. Most plausibly, *prima facie* uncorrelated random processes at individual synapses would result in a stochastic rather than a coherent network of interacting neurons (Hepp 1999). Although Beck (2001) has indicated possibilities (such as quantum stochastic resonance) for achieving coherent patterns at the level of assemblies from fundamentally random synaptic processes, this problem has not been resolved so far.

3.4 Punctuating brain determinism with quantum processes?

More generally speaking, most relations among different levels of neural description are only poorly understood. Yet there are some basic features that apply to most of these relations. For instance, higher (aggregated) level activity typically operates on longer timescales than lower level activity. This entails that fast variability at lower levels typically becomes irrelevant for the slower variability at higher levels. Since indeterministic quantum processes are assumed to occur on low levels (e.g. in synaptic processes, as proposed by Beck and Eccles), this is an argument against their implications at higher levels.

Hameroff and Penrose (1996), and also Stapp (Schwartz *et al.* 2005), have speculated about coherent quantum states at higher levels, but their proposals lack the empirical details necessary to evaluate them seriously. Moreover, it has been argued repeatedly that decoherence processes lead to the decay of such coherent states on timescales much shorter than those of neurobiologically relevant processes, so that neural systems essentially behave quasi-classically.[7]

Since higher-level neural activity has been found to exhibit deterministic chaos, i.e. a sensitive dependence on initial conditions, in particular situations, it has been proposed to consider quantum states as such initial conditions. The idea is that a small deviation due to an indeterminate quantum state could make a (huge) difference for the future course of neural activity, even on longer timescales. Does this idea work?

---

[7] Recent work on particular mechanisms of 'recoherence' (Hartmann *et al.* 2006; Li and Paraoanu 2009), re-establishing coherence in the presence of decohering interactions with a warm and noisy environment, might hold interesting new aspects in this respect, worth exploring in more detail.

First, quantum states playing the role of perturbations of initial conditions introduce limitations on predictability (specifiable by the Kolmogorov-Sinai entropy of the relevant attractor) but the chaotic dynamics itself is classically deterministic. Second, initial perturbations are amplified exponentially only on the attractor, so that its size represents an additional limit for long-time state changes. (Perturbations large enough to push a system out of its attractor are possible but unlikely for quantum fluctuations as expected in the brain.)

Another possibility for faint quantum states to introduce effective changes arises at intrinsically unstable states, for instance at bifurcation points. If a system happens to be at an unstable point, say, at the boundary between two attractors, any arbitrarily small fluctuation decides which of the two attractors the state of the system will relax into. However, the efficacy of small fluctuations in this sense is naturally weakened in the presence of additional, typically classical noise. So again, although possible in principle, it is hard to conceive a scenario in which quantum stochasticity becomes relevant for brain activity at a mesoscopic level correlated with decisions and actions.

Assuming the in-principle possibility that quantum states may make a difference, would this contribute to a better understanding of decision processes? Punctuating brain determinism by quantum stochasticity leads to stochastic changes at best, and this is certainly not how an active decision or action is understood. If quantum stochasticity were relevant only at a micro-level of brain activity, it is easy to argue that it becomes irrelevant (averaged out) at the mesoscopic assembly level. If quantum stochasticity were efficacious *directly* at the mesoscopic level, this would change the picture considerably – but at present this seems doubtful because rapid decoherence would destroy quantum effects too fast to be relevant for the comparably slow timescales of the assembly dynamics.

4. Conclusions

The intricate relations between determinacy and stochasticity raise strong doubts concerning inferences from neurobiological descriptions to ontological statements about the extent of determinism in the brain. These doubts are amplified by the observation that even the kind of description that is suitable depends on the level of brain activity one is interested in. As we move from microscopic (subcellular, membrane-bound molecules) to mesoscopic (neuronal assemblies) and macroscopic (large networks or populations of neurons) levels, completely different kinds of deterministic and stochastic models are suitable and relevant.

These difficulties abound for neurobiological descriptions of the brain as a classical system. So what about quantum states, genuinely (namely ontologically) stochastic states that must be assumed to occur in the brain as they occur everywhere else? As we have shown above, the difficulties of how to understand relations between levels of brain activity become even more pressing, in part because quantum noise is likely exceeded by stronger classical noise, and in part because it is expected only in particular low-level neurobiological processes. It is not overly deprecatory to assess the situation as largely inscrutable.

If, on the other hand, quantum effects are regarded as entry points for any kind of indeterminate or indeterministic influences in an otherwise deterministic system, everything depends on the nature of these influences. Assuming non-physical, e.g. mental, forces acting on neural activity (as proposed by Beck and Eccles 1992) leads into completely unexplored territory – to say the least.

All these (and more) complications entail that well-founded arguments to defend the position that neurobiology 'is' deterministic, or that it 'is' not, are hardly available. Our bottom line is that pretentious claims as to deterministic or indeterministic brain activity are unfounded, and so are the consequences drawn from them. This bears significance with respect to all kinds of problems related to decisions and actions – which are beyond the scope of this article though.

References

Amit, D.J. (1989) *Modeling Brain Function: The World of Attractor Neural Networks*, Cambridge: Cambridge University Press.

Atmanspacher, H. (2000) Ontic and epistemic descriptions of chaotic systems. In D. Dubois (ed.) *Computing Anticipatory Systems CASYS 99*, Berlin: Springer, pp. 465–78.

Atmanspacher, H. and Primas, H. (2003) Epistemic and ontic quantum realities. In L. Castell and O. Ischebeck (eds) *Time, Quantum, and Information*, Berlin: Springer, pp. 301–21.

Atmanspacher, H. and Rotter, S. (2008) Interpreting neurodynamics: concepts and facts, *Cognitive Neurodynamics*, 2: 297–318.

Azouz, R. and Gray, C.M. (1999) Cellular mechanisms contributing to response variability of cortical neurons *in vivo*, *Journal of Neuroscience*, 19: 2209–23.

Beck, F. (2001) Quantum brain dynamics and consciousness. In P. van Loocke (ed.) *The Physical Nature of Consciousness*, Amsterdam: Benjamins, pp. 83–116.

Beck, F. and Eccles, J. (1992) Quantum aspects of brain activity and the role of consciousness, *Proceedings of the National Academy of Sciences of the USA*, 89: 11357–61.

Brunel, N. and Hakim, V. (1999) Fast global oscillations in networks of integrate-and-fire neurons with low firing rates, *Neural Computation*, 11: 1621–71.

Bryant, H.L. and Segundo, J.P. (1976) Spike initiation by transmembrane current: a white-noise analysis, *Journal of Physiology* (London), 260: 279–314.

Freeman, W. (1979) Nonlinear dynamics of pleocortex manifested in the olfactory EEG, *Biological Cybernetics*, 35: 21–34.

Gerstein, G.L. and Mandelbrot, B. (1964) Random walk models for the spike activity of a single neuron, *Biophysical Journal*, 4: 41–68.

Gerstner, W., Kempter, R., van Hemmen, J.L. and Wagner, H. (1996) A neuronal learning rule for sub-millisecond temporal coding, *Nature* 386: 76–8.

Glimcher, P.W. (2005) Indeterminacy in brain and behaviour, *Annual Review of Psychology*, 56: 25–56.

Gustafson, K. (1997) *Lectures on Computational Fluid Dynamics, Mathematical Physics, and Linear Algebra*, Singapore: World Scientific.

Hameroff, S.R. and Penrose, R. (1996) Conscious events as orchestrated spacetime selections, *Journal of Consciousness Studies*, 3(1): 36–53.

Hartmann, L., Dür, W. and Briegel, H.-J. (2006) Steady state entanglement in open and noisy quantum systems at high temperature, *Physical Review A*, 74, 052304.

Hebb, D.O. (1949) *The Organization of Behavior*, Mahwah, NJ: Lawrence Erlbaum.

Hepp, K. (1999) Toward the demolition of a computational quantum brain. In P. Blanchard and A. Jadczyk (eds) *Quantum Future*, Berlin: Springer, pp. 92–104.

Hodgkin, A.L. and Huxley, A.F. (1952) A quantitative description of membrane current and its application to conduction and excitation in nerve, *Journal of Physiology* (London), 117: 500–44.

Jahnke, S., Memmesheimer, R.M. and Timme, M. (2009) How chaotic is the balanced state? *Frontiers in Computational Neuroscience*, 3: 13.

Jammer, M. (1974) *The Philosophy of Quantum Mechanics*, New York: Wiley.

Kaneko, K. (1990) Supertransients, spatiotemporal intermittency, and stability of fully developed spatiotemporal chaos, *Physics Letters A*, 149: 105–12.

Li, J. and Paraoanu, G.S. (2009) Generation and propagation of entanglement in driven coupled-qubit systems, *New Journal of Physics*, 11, 113020.

Mainen, Z.F. and Sejnowski, T.J. (1995) Reliability of spike timing in neocortical neurons, *Science*, 268: 1503–6.

Markram, H., Lübke, J., Frotscher, M. and Sakmann, B. (1997) Regulation of synaptic efficacy by coincidence of postsynaptic APs and EPSPs, *Science*, 275: 213–15.

Sakmann, B. and Neher, E. (1995) *Single-Channel Recording*, New York: Plenum Press.

Scheibe, E. (1974) *The Logical Analysis of Quantum Mechanics*, Oxford: Pergamon.

Schrödinger, E. (1935) Die gegenwärtige Situation in der Quantenmechanik, *Naturwissenschaften*, 23: 807–12, 823–8, 844–9.

Schwartz, J.M., Stapp, H.P. and Beauregard, M. (2005) Quantum theory in neuroscience and psychology: a neurophysical model of mind/brain interaction, *Philosophical Transactions of the Royal Society B*, 360: 1309–27.

Shadlen, M.N. and Newsome, W.T. (1998) The variable discharge of cortical neurons: implications for connectivity, computation, and information coding, *Journal of Neuroscience*, 18: 3870–96.

Softky, W.R. and Koch, C. (1993) The highly irregular firing of cortical cells is consistent with temporal integration of random EPSPs, *Journal of Neuroscience*, 13: 334–50.

Tolhurst, D.J., Movshon, J.A. and Dean, A.F. (1981) The statistical reliability of signals in single neurons in cat and monkey visual cortex, *Vision Research*, 23: 775–85.

Tsuda, I. (2001) Toward an interpretation of dynamic neural activity in terms of chaotic dynamical systems, *Behavioral and Brain Science*, 24: 793–847.

Tuckwell, H.C. (1988) *Introduction to Theoretical Neurobiology*, Vol. 2, Cambridge: Cambridge University Press.

Van Vreeswijk, C. and Sompolinsky, H. (1996) Chaos in neural networks with balanced excitatory and inhibitory activity, *Science*, 274: 1724–6.

Zumdieck, A., Timme, M., Geisel, T. and Wolf, F. (2004) Long chaotic transients in coupled networks, *Physical Review Letters*, 93, 244103.

6

Gödel's incompleteness theorems, free will and mathematical thought

SOLOMON FEFERMAN   *In memory of Torkel Franzén*

1. Logic, determinism and free will

The determinism–free will debate is perhaps as old as philosophy itself and has been engaged in from a great variety of points of view including those of scientific, theological and logical character; my concern here is to limit attention to two arguments from logic. To begin with, there is an argument in support of determinism that dates back to Aristotle, if not farther. It rests on acceptance of the Law of Excluded Middle, according to which every proposition is either true or false, no matter whether the proposition is about the past, present or future. In particular, the argument goes, whatever one does or does not do in the future is determined in the present by the truth or falsity of the corresponding proposition. Surely no such argument could really establish determinism, but one is hard pressed to explain where it goes wrong. One now classic dismantling of it has been given by Gilbert Ryle, in the chapter 'What was to be' of his fine book, *Dilemmas* (Ryle 1954). We leave it to the interested reader to pursue that and the subsequent literature.

The second argument coming from logic is much more modern and sophisticated; it appeals to Gödel's incompleteness theorems (Gödel 1931) to make the case against determinism and in favour of free will, insofar as that applies to the mathematical potentialities of human beings. The claim more precisely is that as a consequence of the incompleteness theorems, those potentialities cannot be exactly circumscribed by the output of any computing machine even allowing unlimited time and space for its work. Here there are several notable proponents, including Gödel himself (with caveats), J.R. Lucas and Roger Penrose. All of these arguments have been subject to considerable critical analysis; it is my purpose here to give some idea of the nature of the claims and debates, concluding with some new considerations that may be in favour of a partial mechanist account of the mathematical mind. Before getting into all that we must first give some explanation both of Gödel's theorems and of the idealized machines due to Alan Turing which connect the formal systems that are the subject of the incompleteness theorems with mechanism.

2. Gödel's incompleteness theorems

The incompleteness theorems concern formal axiomatic systems for various parts of mathematics. The reader is no doubt familiar with one form or another of Euclid's axioms for geometry. Those were long considered to be a model of rigorous logical reasoning from first principles. However, it came to be recognized in the nineteenth century that Euclid's presentation had a number of subtle flaws and gaps, and that led to a much more rigorous presentation of an axiomatic foundation for geometry by David Hilbert in 1899. Hilbert was then emerging as one of the foremost mathematicians of the time, a position he was to hold well into the twentieth century. Axiom systems had also been proposed in the late nineteenth century for other mathematical concepts including the arithmetic of the positive integers by Giuseppe Peano and of the real numbers by Richard Dedekind. In the early twentieth century further very important axiom systems were provided by Ernst Zermelo for sets and by Alfred North Whitehead and Bertrand Russell for a proposed logical foundation of mathematics in their massive work, *Principia Mathematica*. Hilbert recognized that these various axiom systems when fully formally specified could themselves be the subject of mathematical study, for example concerning questions of their consistency, completeness and mutual independence of their constitutive axioms.

As currently explained, a specification of a formal axiom system S is given by a specification of its underlying formal language L and the axioms and rules of inference of S. To set up the language L we must prescribe its basic symbols and then say which finite sequences of basic symbols constitute meaningful expressions of the language; moreover, that is to be done in a way that can be effectively checked, i.e. by a finite algorithmic procedure. The sentences ('closed formulas') of L are singled out among its meaningful expressions; they are generated in an effective way from its basic relations by means of the logical operations. If A is a sentence of L and a definite interpretation of the basic relations of L is given in some domain of objects D then A is true or false under that interpretation. The axioms of S are sentences of L and the rules of inference lead from such sentences to new sentences; again, we need to specify these in a way that can be effectively checked. A sentence of L is said to be *provable in* S if it is the last sentence in a *proof from* S, i.e. a finite sequence of sentences each of which is either an axiom or follows from earlier sentences in the sequence by one of the rules of inference. S is *consistent* if there is no sentence A of L such that both A and its negation (not-A) are provable in S. One of the consequences of Gödel's theorems is that there are formal systems S in the language of arithmetic for which S is consistent yet S proves some sentence A which is false in the domain D of positive integers (1, 2, 3,. . .).

Hilbert introduced the term *metamathematics* for the mathematical study of formal systems for various branches of mathematics. In particular, he proposed as the main program of metamathematics the task of proving the consistency of successively stronger systems of mathematics such as those mentioned above, beginning with the system PA for Peano's Axioms. In order to avoid circularities, Hilbert's program included the proviso that such consistency proofs were to be carried out by the most restrictive mathematical means possible, called *finitistic* by him.

In an effort to carry out Hilbert's program for a substantial part of the formal system PM of *Principia Mathematica* Gödel met a problem which he recognized could turn into a fundamental obstacle for the program. He then recognized that that problem was already met with the system PA. This led to Gödel's stunning theorem that one cannot prove its consistency by any means that can be represented formally within PA, assuming the consistency of PA. In fact, he showed that if S is any formal system which contains PA either directly or via some translation (as is the case with the theory of sets), and if S is consistent, then the consistency of S cannot be proved by any means that can be carried out within S. This is what is called Gödel's *second incompleteness theorem* or his *theorem on the unprovability of consistency*. The *first incompleteness theorem* was the main way-station to its proof; we take it here in the form that if a formal system S is a consistent extension of PA then there is an arithmetical sentence G which is true but not provable in S, where truth here refers to the standard interpretation of the language of PA in the positive integers. That sentence G (called the Gödel sentence for S) expresses of itself that it is not provable in S.

3. Proofs of the incompleteness theorems

We need to say a bit more about how all this works in order to connect Gödel's theorems with Turing machines. It is not possible to go into full detail about how Gödel's theorems are established, but the interested reader will find that there are now a number of excellent expositions at various levels of accessibility which may be consulted for further elaboration.[1] In order to show for these theorems how various metamathematical notions such as provability, consistency, and so on can be expressed in the language of arithmetic, Gödel attached numbers to each symbol in the formal language L of S and then – by using standard techniques for coding finite sequences of numbers by numbers – attached numbers as code to each expression E of L, considered as a finite sequence of basic symbols. These are now called the *Gödel number* of the expression E. In particular, each sentence A of L has a Gödel number. Proofs in S are finite sequences of sentences, and so they too can be given Gödel numbers. Gödel then showed that the Proof-in-S relation, 'n is the number of a proof of the sentence with Gödel number m in S', is definable in the language of arithmetic. Hence if A is a sentence of S and m is its Gödel number then the sentence which says there exists an n such that the Proof-in-S relation holds between n and m expresses that A is provable from S. So we can also express directly from this that A is *not* provable from S. Next, Gödel used an adaptation of what is called *the diagonal method* to construct a specific sentence G, such that PA proves G is equivalent to the sentence expressing that G is not provable in S. Finally, he showed:

(*) If S is consistent then G is (indeed) not provable from S.

It should be clear from the preceding that the statement that S is consistent, i.e. that there is no A such that both A and not-A are provable in S, can also be expressed in the language of arithmetic; we use Con(S) to denote this statement.

---

[1] In particular, I would recommend Franzén (2004) for an introduction at a general level, and Franzén (2005) and Smith (2007) for readers with some background in higher mathematics.

3.1 The second incompleteness theorem (unprovability of consistency)

> If S is a formal system such that S includes PA, and S is consistent, then the sentence Con(S) expressing the consistency of S in arithmetic is not provable in S.

The way Gödel established this is by formalizing the entire argument leading to (*) in Peano Arithmetic. And since the sentence expressing that G is not provable in S is equivalent to G itself, it follows that PA proves:

(**) If Con(S) then G.

So if S were to prove its own consistency statement Con(S) it would also prove G, contrary to (*).

Gödel obtained these remarkable theorems at age 24 as a graduate student at the University of Vienna. The significance of the second incompleteness theorem for Hilbert's program is that if S is a consistent system in which all finitistic methods can be formalized then one cannot give a finitistic consistency proof of S. It was conjectured by John von Neumann that all finitistic methods can be formalized in PA and hence that Hilbert's program would already meet a fundamental obstacle at that point. Gödel did not accept von Neumann's conjecture at first but came around to it within a few years and that is now the common point of view. On the other hand, Hilbert apparently never accepted that Gödel's theorem doomed his consistency program to failure.

4. Turing machines and formal systems

Early in the 1930s, two proposals were made by the logicians Alonzo Church and Jacques Herbrand, respectively, to define the concept of *effective computation procedure* in precise mathematical terms. Gödel found a defect in Herbrand's definition and then modified it so as to avoid its problem. It was then shown by Church and his students that his definition and that of Herbrand-Gödel lead to the same class of computable functions; even so, Gödel did not find either proposal convincing. A couple of years later, the young Cambridge mathematician Alan Turing came up with still another definition in terms of computability on machines of an idealized kind, since then called *Turing machines*. In his paper, Turing (1937) also showed the equivalence of his computability notion with those of Church and Herbrand-Gödel. Church quickly accepted Turing's explication of the informal notion of effective computation procedure as being the most convincing of the three then on offer. Gödel apparently did so too, but the first statement by him in print to that effect was not made until almost thirty years later. That was in a postscript he added to the reprinting of lectures that he had given in Princeton in 1934 in the collection *The Undecidable* (Davis 1965):

> Turing's work gives an analysis of the concept of 'mechanical procedure' (alias 'algorithm' or 'computation procedure' or 'finite combinatorial procedure'). This concept is shown to be equivalent with that of a 'Turing machine'. *A formal system can simply be defined to be any mechanical procedure for producing formulas, called provable formulas.* For any formal system in this sense there exists one in the [usual] sense . . . that has the same provable formulas (and likewise vice versa) . . .
>
> (Gödel 1965, in Gödel 1986: 369, my italics)

Turing's idea was to isolate the most primitive steps of what human computors actually do. The computational work of following a finite set of rules is that of entering (or erasing) a specified list of symbols in various locations and moving from one location to the next. The amount of space and time needed for carrying out a given computation cannot be fixed in advance. Turing reduced this to working on a (potentially) infinite tape divided into a series of squares, of which at any stage of the computation only a finite number are marked. At any active stage in the computation procedure, exactly one instruction is being followed and exactly one square is being scanned; it may be empty or be marked with one of the symbols. The possible actions for a given instruction, noting the state of the scanned square, are to enter a specified symbol or erase its contents, move right or left and proceed to another instruction. (If one is at the left end of the tape, the instruction to move left has no effect.) Beginning with any initial configuration starting at the leftmost square the computation terminates – if at all – when one arrives at an instruction that is designated as the final one. A Turing machine M is specified by its instruction set.

The most primitive alphabet for such computations consists of one symbol, the tally |; each positive integer is then represented by a finite sequence of tallies | |...|, successively marked off on the tape, directly preceded and followed by empty squares. To compute a function f of positive integers such as squaring, i.e. $f(n) = n^2$, one enters n tallies as the initial configuration; the computation is to terminate when it is scanning the rightmost tally of a sequence of n^2 tallies. In general, a function f from positive integers to positive integers is said to be *effectively computable by a Turing machine M* if for each input n as initial configuration the procedure terminates with f(n) as output. By an *effectively enumerable set* of positive integers is meant the range $\{f(1), f(2), f(3),...\}$ of an effectively computable function; there may be repetitions in this range so that it is in fact a finite set. A set is *effectively decidable* if the function f(n) = 1 if n is a member of the set and f(n) = 2 if it is not (called the characteristic function of the set) is effectively computable. Every non-empty effectively decidable set is effectively enumerable, but Turing showed there are effectively enumerable sets that are not effectively decidable. The notion of effectively computable function is extended in a direct way to those with two or more arguments; a relation between two or more arguments is then effectively decidable if its characteristic function is computable.

Given these definitions, Gödel's above stated identification of the most general notion of formal system with 'mechanical procedures for producing formulas' may be spelled out as follows. First of all, given a formal system S, one replaces each formula of the language of S by its Gödel number. By the effectiveness conditions on the specification of S, the set of axioms of S form an effectively decidable set, and each rule of inference, considered as a relation between one or more hypotheses and a conclusion, is an effectively decidable relation between the formulas in each place. It is then an exercise to show that the Proof-in-S relation is effectively decidable. Now define a function f as follows: if n codes a finite sequence whose last term is m, and n is in the Proof-in-S relation to m, then f(n) = m; otherwise, f(n) is the number of some fixed provable sentence selected in advance. Thus the range of f is exactly the set of (Gödel numbers of) provable formulas of S. In terms of the quote from Gödel above, this shows that *given any formal system there is an associated mechanical procedure for producing its provable formulas.*

Conversely, given a formal language L and a Turing machine M for computing some function f, form the set $\{f(1), f(2), f(3),...\}$ and then successively eliminate all terms that are not numbers of sentences in L; the result is still effectively enumerable, and its set of purely logical consequences is the set of provable formulas of a suitable formal system S in the language L. Thus *any mechanical procedure may be effectively transformed into another such procedure for producing the provable formulas of a formal system.*

Given Gödel's identification in these senses of formal systems with mechanical procedures, one is led to the following formulation of the thesis that the mathematical mind is mechanical:

The Formalist-Mechanist Thesis I

Insofar as human mathematical thought is concerned, mind is mechanical in that the set of all mathematical theorems, actual or potential, is the set of provable sentences of some formal system.

Note well that this is a thesis concerning mathematical thought only. Of course that would be a consequence of all mental activity being determined in some way by a machine. But the thesis is compatible with thought in general not being describable in mechanistic terms. We shall abbreviate this thesis as FMT I.

In the following we shall concentrate on two thinkers who deny FMT I to some extent or other, namely Gödel and Lucas.

5. Gödel on minds and machines

Gödel first laid out his thoughts in this direction in what is usually referred to as his 1951 Gibbs lecture, 'Some basic theorems on the foundations of mathematics and their implications' (Gödel 1951).[2] The text of this lecture was never published in his lifetime, though he wrote of his intention to do so soon after delivering it. After Gödel died, it languished with a number of other important essays and lectures in his *Nachlass* until it was retrieved for publication in Volume III of Gödel's *Collected Works* (1995: 304–23).

There are essentially two parts to the Gibbs lecture, both drawing conclusions from the incompleteness theorems. The first part concerns the potentialities of mind vs. machines for the discovery of mathematical truths. The second part is an argument aimed to 'disprove the view that mathematics is only our own creation', and thus to support some version of platonic realism in mathematics; only the first part concerns us here.[3] Gödel there highlighted the following dichotomy:

---

[2] Gödel's lecture was the twenty-fifth in a distinguished series set up by the American Mathematical Society to honour the nineteenth-century American mathematician, Josiah Willard Gibbs, famous for his contributions to both pure and applied mathematics. It was delivered to a meeting of the AMS held at Brown University on 26 December 1951.

[3] George Boolos wrote a very useful introductory note to both parts of the Gibbs lecture in Vol. III of the Gödel *Works*. More recently I have published an extensive critical analysis of the first part, under the title 'Are there absolutely unsolvable problems? Gödel's dichotomy' (Feferman 2006), followed by the closely related 'Gödel, Nagel, minds and machines' (Feferman 2009), on both of which I draw extensively in the following.

> *Either . . . the human mind (even within the realm of pure mathematics) infinitely*
> *surpasses the powers of any finite machine, or else there exist absolutely unsolvable*
> *diophantine problems . . .*
>
> (Gödel 1995: 310)

By a *diophantine problem* is meant a proposition of the language of Peano Arithmetic of a relatively simple form whose truth or falsity is to be determined; its exact description is not important to us.[4] Gödel showed that the consistency of a formal system is equivalent to a diophantine problem, to begin with by expressing it in the form that no number codes a proof of a contradiction. According to Gödel, his dichotomy is a 'mathematically established fact' which is a consequence of the incompleteness theorem. However, all that he says by way of an argument for it is the following:

> [I]f the human mind were equivalent to a finite machine then objective
> mathematics not only would be incompletable in the sense of not being
> contained in any well-defined axiomatic system, but moreover there would
> exist *absolutely* unsolvable problems . . . where the epithet 'absolutely' means
> that they would be undecidable, not just within some particular axiomatic
> system, but by *any* mathematical proof the mind can conceive.
>
> (Gödel 1995: 310, italics in original)

By a *finite machine* here Gödel means a Turing machine, and by a *well-defined axiomatic system* he means an effectively specified formal system; as explained above, he takes these to be equivalent in the sense that the set of theorems provable in such a system is the same as the set of theorems that can be effectively enumerated by such a machine. Thus, to say that the human mind is equivalent to a finite machine 'even within the realm of pure mathematics' is another way of saying that what the human mind can *in principle* demonstrate in mathematics is the same as the set of theorems of some formal system, i.e. that FMT I holds. By *objective mathematics* Gödel means the totality of true statements of mathematics, which includes the totality of true statements of arithmetic. Then the assertion that objective mathematics is incompletable is simply a consequence of the second incompleteness theorem.

Examined more closely, Gödel's argument is that if the human mind were equivalent to a finite machine, or – what comes to the same thing –

---

[4] Nowadays, mathematicians reserve the terminology 'diophantine equations' and 'diophantine problems' to a more specialized class than taken by Gödel. However, Gödel's have been shown to be equivalent to the non-existence of solutions to suitable diophantine equations.

an effectively presented formal system S, then there would be *a true statement that could never be humanly proved,* namely Con(S). So that statement would be *absolutely undecidable* by the human mind, and moreover it would be equivalent to a diophantine statement. *Note however, the tacit assumption that the human mind is consistent*; otherwise, it is equivalent to a formal system in a trivial way, namely one that proves all statements. Actually, Gödel apparently accepts a much stronger assumption, namely that we prove *only* true statements; but for his argument, only the weaker assumption is necessary (together of course with the assumption that PA or some comparable basic system of arithmetic to which the second incompleteness theorem applies has been humanly accepted).

Though he took care to formulate the possibility that the second term of the disjunction holds, there's a lot of evidence outside of the Gibbs lecture that Gödel was convinced of the anti-mechanist position as expressed in the first disjunct. That's supplied, for example, in his informal communication of various ideas about minds and machines to Hao Wang, initially in the book, *From Mathematics to Philosophy* (Wang 1974: 324–6), and then at greater length in *A Logical Journey: From Gödel to Philosophy* (Wang 1996, especially ch. 6). So why didn't Gödel state that outright in the Gibbs lecture instead of the more cautious disjunction in the dichotomy? The reason was simply that he did not have an unassailable proof of the falsity of the mechanist position. Indeed, despite his views concerning the 'impossibility of physico-chemical explanations of . . . human reason' he raised some caveats in a series of three footnotes to the Gibbs lecture, the second of which is as follows:

> [I]t is conceivable . . . that brain physiology would advance so far that it would be known with empirical certainty
>
> 1. that the brain suffices for the explanation of all mental phenomena and is a machine in the sense of Turing;
>
> 2. that such and such is the precise anatomical structure and physiological functioning of the part of the brain which performs mathematical thinking.
> (Gödel 1995: 310)

Some twenty years later, Georg Kreisel made a similar point in terms of formal systems rather than Turing machines:

> [I]t has been clear since Gödel's discovery of the incompleteness of formal systems that we could not have *mathematical* evidence for the adequacy of any formal system; but this does not refute the possibility that some quite specific system . . . encompasses all possibilities of (correct) mathematical reasoning . . .

> *In fact the possibility is to be considered that we have some kind of nonmathematical evidence for the adequacy of such* [a system].
>
> (Kreisel 1972: 322, my italics)

I shall call the genuine possibility entertained by Gödel and Kreisel, *the mechanist's empirical defence* (or *escape hatch*) against claims to have *proved* that mind exceeds mechanism on the basis of the incompleteness theorems, that is that FMT I is wrong.

6. Lucas on minds and machines

The first outright such claim was made by the Oxford philosopher J.R. Lucas in his article, 'Minds, machines and Gödel': 'Gödel's theorem seems to me to prove that Mechanism is false, that is, that minds cannot be explained as machines' (Lucas 1961: 112). His argument is to suppose that there is a candidate machine M (called by him a 'cybernetical machine') that enumerates exactly the mathematical sentences that can be established to be true by the human mind, hence exactly what can be proved in a formal system for humanly provable truths. Assuming that,

> [we] now construct a Gödelian formula [the sentence G described in Section 3 above] in this formal system. This formula cannot be *proved-in-the-system*. Therefore the machine cannot produce the corresponding formula as being true. But we can see that the Gödelian formula is true: any rational being could follow Gödel's argument, and convince himself that the Gödelian formula, although unprovable-in-the-system, was nonetheless . . . true. . . . This shows that a machine cannot be a complete and adequate model of the mind. It cannot do *everything* that a mind can do, since however much it can do, there is always something which it cannot do, and a mind can. . . . therefore we cannot hope ever to produce a machine that will be able to do all that a mind can do: we can never not even in principle, have a mechanical model of the mind.
>
> (Lucas 1961: 115, italics in original)

Paul Benacerraf and Hilary Putnam soon objected to Lucas's argument on the grounds that he was assuming it is known that one's mind is consistent, since Gödel's theorem only applies to consistent formal systems. But Lucas had already addressed this as follows:

> a mind, *if it were really a machine*, could not reach the conclusion that it was a consistent one. [But] for a mind which is not a machine no such conclusion follows. . . . It therefore seems to me both proper and reasonable for a mind

to assert its own consistency: proper, because although machines, as we might have expected, are unable to reflect fully on their own performance and powers, yet to be able to be self-conscious in this way is just what we expect of minds: and reasonable, for the reasons given. Not only can we fairly say simply that we *know* we are consistent, apart from our mistakes, but we must in any case *assume* that we are, if thought is to be possible at all; . . . and finally we can, in a sense, *decide* to be consistent, in the sense that we can resolve not to tolerate inconsistencies in our thinking and speaking, and to eliminate them, if ever they should appear, by withdrawing and cancelling one limb of the contradiction.

(Lucas 1961: 124, italics in original)

In this last, there is a whiff of the assertion of human free will. Lucas is more explicit about the connection in the conclusion to his essay:

If the proof of the falsity of mechanism is valid, it is of the greatest consequence for the whole of philosophy. Since the time of Newton, the bogey of mechanist determinism has obsessed philosophers. If we were to be scientific, it seemed that we must look on human beings as determined automata, and not as autonomous moral agents . . . But now, though many arguments against human freedom still remain, the argument from mechanism, perhaps the most compelling argument of them all, has lost its power. No longer on this count will it be incumbent on the natural philosopher to deny freedom in the name of science: no longer will the moralist feel the urge to abolish knowledge to make room for faith. We can even begin to see how there could be room for morality, without its being necessary to abolish or even to circumscribe the province of science. Our argument has set no limits to scientific enquiry: it will still be possible to investigate the working of the brain. It will still be possible to produce mechanical models of the mind. Only, now we can see that no mechanical model will be completely adequate, nor any explanations in purely mechanist terms. We can produce models and explanations, and they will be illuminating: but, however far they go, there will always remain more to be said. There is no arbitrary bound to scientific enquiry: but no scientific enquiry can ever exhaust the infinite variety of the human mind.

(Lucas 1961: 127)

According to Lucas, then, FMT I is *in principle* false, though there can be scientific evidence for the mechanical workings of the mind to some extent or other insofar as mathematics is concerned. What his arguments do not countenance is the possibility of obtaining fully convincing empirical support for the mechanist thesis, namely that eventually all evidence points to mind being mechanical though we cannot ever hope to supply a *complete perfect description* of a formal system which accounts for its workings.[5] Moreover, such a putative system need not necessarily be consistent. Without such a perfect description for a consistent system as a model of the mind, the argument for Gödel's theorem cannot apply. Lucas, in response to such a suggestion has tried to shift the burden to the mechanist: 'The consistency of the machine is established not by the mathematical ability of the mind, but on the word of the mechanist' (Lucas 1996), a burden that the mechanist can refuse to shoulder by simply citing this empirical defence. Finally, the compatibility of FMT I with a non-mechanistic account for thought in general would still leave an enormous amount of room for morality and the exercise of free will.

Despite such criticisms, Lucas has stoutly defended to the present day his case against the mechanist on Gödelian grounds. One can find on his home page[6] most of his published rejoinders to various of these as well as further useful references to the debate. The above quotations do not by any means exhaust the claims and arguments in his thoroughly thought-out discussions.

7. Critiques of Gödelian arguments against mechanism[7]

Roger Penrose is another well-known defender of the Gödelian basis for anti-mechanism, most notably in his two books, *The Emperor's New Mind* (1989), and *Shadows of the Mind* (1994). Sensitive to the objections to Lucas, he claimed in the latter only to have proved something more modest (and in accord with experience) from the incompleteness theorems: 'Human mathematicians are not using a knowably sound algorithm in order to ascertain mathematical truth' (Penrose 1994: 76). But later in that work he came up with a new argument purported to show that the human mathematician can't even consistently *believe* that his mathematical thought is circumscribed by a mechanical algorithm (Penrose 1994, sections 3.16 and 3.23). Extensive critiques have been made of Penrose's original and new arguments in an issue of the journal *Psyche* (1996), to which he responded

---

[5] That would be analogous to obtaining fully convincing empirical support for the thesis that the workings of, say, the human auditory and visual systems are fully explicable in neurological and physical terms, though one will never be able to produce a complete perfect description of how those operate. I presume that we are in fact in such a position.

[6] http://users.ox.ac.uk/~jrlucas/

[7] This section is drawn directly from Feferman (2009).

in the same issue. And more recently, Stewart Shapiro (2003) and Per Lindström (2001, 2006) have carefully analysed and critiqued his 'new argument'. But Penrose has continued to defend it, as he did in his public lecture for the Gödel Centenary Conference held in Vienna in April 2006.

Historically, there are many examples of mathematical proofs of what can't be done in mathematics by specific procedures, e.g. the squaring of the circle, or the solution by radicals of the quintic, or the solvability of the halting problem. But it is hubris to think that by mathematics alone we can determine what the mathematician can or cannot do in general. The claims by Gödel, Lucas and Penrose to do just that from the incompleteness theorems depend on making highly idealized assumptions both about the nature of mind and the nature of machines. A very useful critical examination of these claims and the underlying assumptions has been made by Shapiro in his article, 'Incompleteness, mechanism and optimism' (1998), including the following points. First of all, how are we to understand the mathematizing capacity of the human mind, since what is at issue is the producibility of an infinite set of propositions? No one mathematician, whose life is finitely limited, can produce such a list, so either what one is talking about is what the individual mathematician *could do in principle*, or we are talking in some sense about the potentialities of the pooled efforts of the community of mathematicians now or ever to exist. But even that must be regarded as a matter of what can be done *in principle*, since it is most likely that the human race will eventually be wiped out either by natural causes or through its own self-destructive tendencies by the time the sun ceases to support life on earth.

What about the assumption that the human mind is consistent? In practice, mathematicians certainly make errors and thence arrive at false conclusions that in some cases go long undetected. Penrose, among others, has pointed out that when errors are detected, mathematicians seek out their source and correct them (see Penrose 1996: 137ff.), and so he has argued that it is reasonable to ascribe self-correctability and hence consistency to our idealized mathematician. But even if such a one can correct all his errors, can he know with mathematical certitude, as required for Gödel's claim, that he is consistent?

As Shapiro points out, the relation of both of these idealizations to practice is analogous to the competence/performance distinction in linguistics.

There are two further points of idealization to be added to those considered by Shapiro. The first of these is the assumption that the notions and statements of mathematics are fully and faithfully expressible in a formal

115

language, so that what can be humanly proved can be compared with what can be the output of a machine. In this respect it is usually pointed out that the only part of the assumption that needs be made is that the notions and statements of elementary number theory are fully and faithfully represented in the language of first-order arithmetic, and that among those only simply universal ('diophantine') statements need be considered, since that is the arithmetized form of the consistency statements for formal systems. But even this idealization requires that statements of unlimited size must be accessible to human comprehension.

Finally to be questioned is the identification of the notion of finite machine with that of Turing machine. Turing's widely accepted explication of the informal concept of effective computability puts no restriction on time or space that might be required to carry out computations. But the point of that idealization was to give the strongest *negative* results, to show that certain kinds of problems can't be decided by a computing machine, no matter how much time and space we allow. And so if we carry the Turing analysis over to the potentiality of mind in its mathematizing capacity, to say that mind infinitely surpasses any finite machine is to say something even stronger. It would be truly impressive if that could be definitively established, but none of the arguments that have been offered are resistant to the mechanist's empirical defence. Moreover, suppose that the mechanist is right, and that in some reasonable sense mind *is* equivalent to a finite machine; is it appropriate to formulate that in terms of the identification of what is humanly provable with what can be enumerated by a Turing machine? Isn't the mechanist aiming at something stronger in the opposite direction, namely an explanation of the mechanisms that govern the production of human proofs?

8. Mechanism and partial freedom of the will

This last point is where I think something new has to be said, something that I have already drawn attention to (in Feferman 2006, 2009). Namely, there is an *equivocation* involved, that lies in identifying *how* the mathematical mind works with the totality of *what* it can prove. Again, the difference is analogous to what is met in the study of natural language, where we are concerned with the *way* in which linguistically correct utterances are generated and *not* with the potential totality of *all* such utterances. That would seem to suggest that if one is to consider *any*

116

idealized formulation of the mechanist's position at all in logical terms, it ought to be of the mind as one *constrained* by the axioms and rules of some effectively presented formal system. Since in following those axioms and rules one has *choices* to be made at each step, *at best* that identifies the mathematizing mind with *the program for a non-deterministic Turing machine*, and *not* with the set of its enumerable statements (even though that can equally well be supplied by a deterministic Turing machine).[8] One could no more disprove this modified version of the idealized mechanist's thesis than the version considered by Gödel, and others, simply by applying the mechanist's empiricist argument. Nevertheless, it is difficult to conceive of any formal system of the sort with which we are familiar, from Peano Arithmetic (PA) up to Zermelo-Fraenkel Set Theory (ZF) and beyond, actually underlying mathematical thought as it is experienced.

As I see it, a principal reason for the implausibility of this modified version of the mechanist's thesis lies in the concept of a formal system S that is currently taken for granted in logical work. An essential part of that concept is that the language L of S is fixed once and for all. For example, the language of PA is determined (in one version) by taking the basic symbols to be those for equality, zero, successor, addition and multiplication, and that of ZF is fixed by taking its basic symbols to be those for equality and membership. This forces axiom schemata that may be used in such systems, such as for mathematical induction in arithmetic and separation in set theory, to be infinite bundles of all possible substitution instances by formulas from that language; this makes metamathematical but not mathematical sense. Besides that, the restriction of mathematical discourse to a language fixed in advance, even if only implicitly, is completely foreign to mathematical practice.

In recent years I have undertaken the development of a modified conception of formal system that does justice to the openness of practice and yet gives it an underlying rule-governed logical-axiomatic structure; it thus suggests a way, admittedly rather speculative, of straddling the Gödelian dichotomy. This is in terms of a notion of *open-ended schematic axiomatic system*, i.e. one whose schemata are finitely specified by means of propositional and predicate variables (thus putting the 'form' back into 'formal systems') while the language of such a system is considered to be *open-ended*, in the sense that its basic vocabulary may be expanded to any wider

---

[8] Lucas (1961: 113–14) recognized the equivalence of non-deterministic and deterministic Turing machines with respect to the set of theorems proved by each.

conceptual context in which its notions and axioms may be appropriately applied. In other words, on this approach, *implicit in the acceptance of given schemata is the acceptance of any meaningful substitution instances that one may come to meet*, but which those instances are is not determined by restriction to a specific language fixed in advance (see Feferman 1996 and 2006a, and Feferman and Strahm 2000). The idea is familiar from logic with such basic principles as 'P and Q implies P' and rules such as 'from P and P implies Q, infer Q', for arbitrary propositions P and Q. But it is directly extended to the principle of mathematical induction for any property P ('if P(1) and for all n, P(n) implies P(n+1), then for all positive integers n, P(n)'), and Zermelo's separation axiom for any property P ('if a is any set then there is a set b such that for all x, x is a member of b if and only if x is a member of a and P(x) holds). All of these may be considered (and are actually employed) without restriction to any specific language fixed in advance.

This leads me to suggest the following revision of FMT I:

The Formalist-Mechanist Thesis II

Insofar as human mathematical thought is concerned, mind is mechanical in that it is completely constrained by some open-ended schematic formal system.

If the concepts of mathematics turned out to be limited to those that can be expressed in one basic formal language L, the two theses would be equivalent. So the point of this second thesis is that the conceptual vocabulary of mathematics is not necessarily limited in that way, but that mathematics is otherwise constrained once and for all by the claimed finite number of open-ended schematic principles and rules. The idea is spelled out in the final section of Feferman (2009), to which the reader is referred, given the limitations of space here. But I will repeat some of the arguments as to why the language of mathematics should be considered to be open-ended, i.e. not restricted to one language L once and for all.

Consider, to the contrary, the claim by many that all mathematical concepts are definable in the language of axiomatic set theory. It is indeed the case that the current concepts of working ('pure') mathematicians are with few exceptions all expressible in set theory. But there are genuine outliers. For example a natural and to all appearances coherent mathematical notion whose full use is not set-theoretically definable is that of a category; only so-called 'small' categories can be directly treated in that way (see Mac Lane 1971 and Feferman 1977 and 2006b). Other outliers are to be found on the constructive fringe of mathematics in the schools of Brouwerian intuitionism and Bishop's constructivism (see Beeson 1985) whose basic notions and principles are not directly accounted for in set theory with its essential use of classical logic. And it may be argued that there are informal mathematical concepts like those of knots, or infinitesimal displacements on a smooth surface, or of random variables, to name just a few, which may be the subject of convincing mathematical reasoning but that are accounted for in set theory only by some substitute notions that share the main expected properties but are not explications in the ordinary sense of the word. Moreover, the idea that set-theoretical concepts and questions like Cantor's continuum problem have determinate mathematical meaning has been challenged on philosophical grounds (Feferman 2000). Finally, there is a theoretical argument for openness, even if one accepts the language L of set theory as a determinately meaningful one. Namely, by Tarski's theorem, the notion of truth T_L for L is not definable in L; and then the notion of truth for the language obtained by adjoining T_L to L is not definable in *that* language, and so on (even into the transfinite).

Another argument that may be made against the restriction of mathematics to a language fixed in advance is historical. Simply witness the progressive amplification of the body of mathematical concepts since the emergence of abstract mathematics in Greek times. It would be hubris to suppose that that process will ever be brought to completion. But having generally granted that certain open-ended schematic principles and rules completely govern all *logical* thinking, it is not hubris to grant the possibility that there are certain (finitely many) open-ended schematic principles and rules that completely constrain all *mathematical* thinking now and ever to come, no matter what new concepts and specifically associated principles one comes to accept. Of course, by the mechanist's empiricist argument one could no more disprove this version FMT II of the mechanist's thesis than the version FMT I considered by Gödel, Lucas, Penrose, and others.[9]

---

[9] Another kind of suggested combination of openness with mechanism that could evade the arguments from Gödel's theorems has recently been brought to my attention by Martin Solomon: '[I]f we treat mechanical theorem recognizers as *open* systems, continually interacting with their environment, they may enjoy increases in power (possibly through "random inspirations" from this environment) which could occur in a "surprising", i.e. non-computable manner' (Lyngzeidetson and Solomon 1994: 552). The idea is that these systems respond to varying input data that may be non-computable in a completely computable way in order to generate mathematical theorems. However, no criterion could be built into such a system to insure that only true statements are proved. My open-ended schematic formal systems are also not immune to that problem if faulty concepts are adopted. For example, the concept of 'feasibly computable number' leads to the conclusion that all numbers are feasible by applying the induction scheme.

That mathematics is constrained by its modes of reasoning in some way or other accords with ordinary experience; that and much work in the formalization of mathematical thought is what gives plausibility to the FMT II thesis. That the practice of mathematics provides extraordinary scope for the exercise of creative free will is also a feature of everyday experience. But that that free will may only be partial in the sense of FMT II need be no more surprising than that the human exercise of free will as applied to bodily actions is constrained by the laws of natural science. I am taking all this in a prima facie sense. Lacking any sort of convincing argument for genuine free will, what is at issue here is whether the laws of nature or thought as we know them leave open the possibility of making real choices at each step in our physical and intellectual lives, in particular in the case of mathematical thought, new choices as to the concepts with which mathematics may deal.

On the other hand, one may well ask to what extent FMT II fits with a mechanistic view of the human mind as a whole.[10] Indeed, one ought to pose that of the stronger thesis FMT I as well, even though Gödel, Lucas and Penrose all considered that a proof of its falsity would amount to a rejection of mechanism, at least in the mental realm. There are actually two competing mechanistic theories of the mind in current cognitive science, the digital computational model identified with such figures as Alan Turing and John von Neumann, and the connectionist computational model exemplified by the work of John McClelland and David Rumelhart; see Harnish (2002) for an excellent historical introduction to and expository survey of these two approaches. It is only the digital computational conception that is the target of the anti-mechanists who argue from Gödel's incompleteness theorems. Though FMT I is a consequence of that viewpoint, the converse does not hold since FMT I only concerns the mathematizing capacities of the human mind. Despite the empirical defence in possible support of it, the evidence for FMT I in mathematical practice is actually practically nil. Nevertheless, what has been at issue here is whether it can be disproved on the basis of Gödel's theorems, and I have argued along with others that it cannot. If that is granted, it is theoretically possible that FMT I holds without that making the case for the digital computational model of the mind as a whole. FMT II is even farther from that point of view but it seems to me to be similar enough to FMT I to warrant being

---

[10] I wish to thank Mr Lucas for raising this question.

called a Formalist-Mechanist Thesis. Speaking for myself, I believe something like FMT II is true but do not subscribe to any mechanistic conception of the mind as a whole.

References

Beeson, M. (1985) *Foundations of Constructive Mathematics*, Berlin: Springer-Verlag.

Davis, M. (ed.) (1965) *The Undecidable: Basic Papers on Undecidable Propositions, Unsolvable Problems and Computable Functions*, Hewlett, NY: Raven Press.

Feferman, S. (1977) Categorical foundations and foundations of Category Theory. In R.E. Butts and J. Hintikka (eds) *Logic, Foundations of Mathematics and Computability Theory*, Vol. I, Dordrecht: Reidel, pp. 149–65.

Feferman, S. (1995) Penrose's Gödelian argument, *Psyche*, 2: 21–32; also at http://psyche.cs.monash.edu.au/v2/psyche-2-07-feferman.html

Feferman, S. (1996) Gödel's program for new axioms: why, where, how and what? In P. Hájek (ed.) *Gödel '96, Lecture Notes in Logic 6*, Berlin: Springer, pp. 3–22.

Feferman, S. (2000) Does mathematics need new axioms? *Bulletin of Symbolic Logic*, 6: 401–13.

Feferman, S. (2006) Are there absolutely unsolvable problems? Gödel's dichotomy, *Philosophia Mathematica*, III, 14: 134–52.

Feferman, S. (2006a) Open-ended schematic axiom systems (abstract), *Bulletin of Symbolic Logic*, 12: 145.

Feferman, S. (2006b) Enriched stratified systems for the foundations of Category Theory. In G. Sica (ed.) *What is Category Theory?*, Monza: Polimetrica, pp. 185–203.

Feferman, S. (2009) Gödel, Nagel, minds and machines, *Journal of Philosophy*, CVI(4): 201–19.

Feferman, S. and T. Strahm (2000) The unfolding of non-finitist arithmetic, *Annals of Pure and Applied Logic*, 104: 75–96.

Franzén, T. (2004) *Inexhaustibility: A Non-exhaustive Treatment*, Wellesley, MA: A.K. Peters.

Franzén, T. (2005) *Gödel's Theorem: An Incomplete Guide to its Use and Abuse*, Wellesley, MA: A.K. Peters.

Gödel, K. (1931) Über Formal Unentscheidbare Sätze der Principia Mathematica und Verwandter Systeme I, *Monatshefte für Mathematik und Physik*, 38: 173–98; reprinted with facing English translation in Gödel (1986), pp. 144–95.

Gödel, K. (1934) On undecidable propositions of formal mathematical systems (mimeographed lecture notes at the Institute for Advanced Study, Princeton, NJ); reprinted in Davis (1965), pp. 39–74 and in Gödel (1986), pp. 346–71.

Gödel, K. (1951) Some basic theorems on the foundations of mathematics and their implications. In Gödel (1995), pp. 304–23.

Gödel, K. (1986) *Collected Works, Vol. I: Publications 1929–1936* (S. Feferman,

J.W. Dawson, Jr., S.C. Kleene, G.H. Moore, R.M. Solovay and J. van Heijenoort, eds), New York: Oxford University Press.

Gödel, K. (1995) *Collected Works, Vol. III: Unpublished Essays and Lectures* (S. Feferman, J.W. Dawson, Jr., W. Goldfarb, C. Parsons and R.M. Solovay, eds), New York: Oxford University Press.

Harnish, R.M. (2002) *Minds, Brains and Computers: An Historical Introduction to the Foundations of Cognitive Science*, Oxford: Blackwell.

Kreisel, G. (1972) Which number-theoretic problems can be solved in recursive progressions on π^1_1 paths through O?, *Journal of Symbolic Logic*, 37: 311–34.

Lindström, P. (2001) Penrose's new argument, *Journal of Philosophical Logic*, 30: 241–50.

Lindström, P. (2006) Remarks on Penrose's 'new argument', *Journal of Philosophical Logic*, 35: 231–7.

Lucas, J.R. (1961) Minds, machines and Gödel, *Philosophy*, 36: 112–37.

Lucas, J.R. (1996) Minds, machines and Gödel: a retrospect. In P.J.R. Millican and A. Clark (eds) *Machines and Thought: The Legacy of Alan Turing*, Vol. 1, Oxford: Oxford University Press, pp. 103–24.

Lyngzeidetson, A.E. and M.K. Solomon (1994) Abstract complexity and the mind-machine problem, *British Journal for the Philosophy of Science*, 45: 549–54.

Mac Lane, S. (1971) *Categories for the Working Mathematician*, Berlin: Springer-Verlag.

Penrose, R. (1989) *The Emperor's New Mind*, Oxford: Oxford University Press.

Penrose, R. (1994) *Shadows of the Mind*, Oxford: Oxford University Press.

Penrose, R. (1996) Beyond the doubting of a shadow, *Psyche*, 2: 89–129; also at http://psyche.cs.monash.edu.au/v2/psyche-2-23-penrose.html.

Ryle, G. (1954) *Dilemmas*, Cambridge: Cambridge University Press.

Shapiro, S. (1998) Incompleteness, mechanism, and optimism, *Bulletin of Symbolic Logic*, 4: 273–302.

Shapiro, S. (2003) Mechanism, truth, and Penrose's new argument, *Journal of Philosophical Logic*, 32: 19–42.

Smith, P. (2007) *An Introduction to Gödel's Theorems*, Cambridge: Cambridge University Press.

Turing, A.M. (1937) On computable numbers, with an application to the Entscheidungsproblem, *Proceedings of the London Mathematics Society* (2), 42: 230–65; correction, ibid. 43: 544–6; reprinted in Davis (1965), pp. 116–54; and in Turing (2001), pp. 18–56.

Turing, A.M. (2001) Collected works of A.M. Turing. In R.O. Gandy and C.E.M. Yates (eds) *Mathematical Logic*, Amsterdam: North-Holland/Elsevier.

Wang, H. (1974) *From Mathematics to Philosophy*, London: Routledge & Kegan Paul.

Wang, H. (1996) *A Logical Journey: From Gödel to Philosophy*, Cambridge, MA: MIT Press.

7
Feferman on Gödel and free will

A response to Chapter 6

J.R. LUCAS

Feferman is right to dismiss logical determinism perfunctorily, although it puzzled Aristotle and the mediaeval Schoolmen and many people still. The inference from 'There is going to be either a sea battle tomorrow or not' to 'Either there is going to be a sea battle tomorrow or there is not going to be a sea battle tomorrow' is invalid for the same reason as the inference from 'I know he is either in his room or in the library' to 'Either I know he is in his room or I know he is in the library'. More generally, 'It is necessary that (either p or q)' does not entail 'Either (it is necessary that p) or (it is necessary that q)'. Like Gilbert Ryle, I spent time working through the argument in detail and delving into tense logic, an interesting topic that need not detain us here.[1]

Feferman gives a careful account of the much-criticized Gödelian argument against mechanism put forward by Gödel, Penrose and me. Like many other critics he highlights our assumption that any plausible mechanical model of the mind must be consistent. Is that assumption justified? By what right do we make out that our understanding of mathematics is consistent? Frege thought his was, and was wrong. How can we be confident that Zermelo-Fraenkel Set Theory (ZF) does not contain a hidden contradiction that may one day emerge?

So far as our understanding of mathematics is concerned, the answer is simple: if ZF turned out to be inconsistent, we should reformulate Set Theory, modifying or abandoning some axiom required for proving the contradiction. There is no guarantee that our current formalizations of

[1] J.R. Lucas, *The Future*, Oxford: Blackwell, 1989.

mathematics are correct, or consistent. But that is not the point. What is in issue is not the consistency of our current formalizations of mathematics, but the consistency of a purported representation of the whole of a person's mental activity, and hence that his mathematical output is a subset of the set of provable sentences of some formal system. The mechanist might argue for this on empirical grounds – the success of neuro-physiology and, until recently, an over-arching Laplacian determinism. If the empiricist argument succeeded, then the whole of a person's mental activity could be represented by a mechanism, and hence the whole of his mathematical work. But the Gödelian argument shows that that cannot be the case, and so not only Feferman's Formalist-Mechanist Thesis I is shown to be false, but the general mechanist thesis too.

Feferman complains that we work with highly idealized views of the nature of mind and of mechanism, and that the empirical support for the mechanist thesis cannot be expected to supply a complete perfect description of the workings of the brain any more than of the digestive system. But the mechanist's 'empirical defence' claims that there is such a system, even though it cannot offer a complete and perfect description of it. And this system, although not completely or perfectly described by the empirical scientist, must be completely and perfectly describable in principle. If this idealized representation of a particular human mind is consistent, then according to the claim it should not be able to do something that he can do. And if the idealized representation is not consistent, then it is not selective, and does not distinguish between truth and falsity, which a human mind at least sometimes can do. So either way the Formalist-Mechanist Thesis I fails.

Many critics fail to distinguish this informal argument from a formal proof of consistency within the idealized representation. 'Merely to find from a given machine M a statement S from which it can be proved that M, if consistent, cannot prove S is not to prove S – even if M is consistent.'[2] But the word 'prove' is being used equivocally, the first two times meaning formally proved by M, the third meaning established as true by cogent argument not necessarily confined to the formal proof-procedures of M. The contention that if M were inconsistent, it would not be selective, and hence not a realistic model of a human mind is a cogent argument, though not a formal proof within M, and thus enables a human mind to prove

[2] George Boolos, Introductory note to the '*1951' chapter, Kurt Gödel, *Collected Works III*, Oxford, 1995, p. 295; reprinted in George Boolos, *Logic, Logic, and Logic*, Cambridge, MA: Harvard University Press, 1998, p. 110.

beyond reasonable doubt, though not by a formal proof-sequence generated by M, that if M is made out to be a realistic model of his mind, there is a truth which he can establish and it cannot.

But perhaps, it might be suggested, the inconsistency is buried very deep, and only emerges after a long time, if at all. A Nietzschean might maintain that philosophers who persist in thinking too long and too hard about tricky concepts, such as the Self and God, would end up by going mad. It is an intelligible hypothesis. I leave it to others to develop and demolish it.

In his last section, Mechanism and partial freedom of the will, Feferman puts forward a mediating position between hardline mechanism and the anti-mechanist position taken by Gödel, Penrose and me. It depends on an 'open-ended schematic axiomatic system', which is open-ended 'in the sense that its basic vocabulary may be expanded to any wider conceptual context in which its notions and axioms may be appropriately applied'. This fits with mathematical practice – witness how the concept of number has been widened from the natural numbers to the integers, the rationals, the reals and the complex numbers – but is difficult to construe within the framework of formal logic. Standardly in logic the range of possible substitution instances is laid down in the initial formulation. The thus circumscribed possibilities are too restricted for many purposes. Feferman's opening up the range of admissible substitution extends these possibilities in an interesting way, and may yield valuable insights into the nature of mathematical thinking – which does, indeed, progress as much by the recognition of concepts as by the discovery of derivations. But even if its axiom schemata and rules of inference are fixed, a system with open-ended rules of substitution is not going to give the hardline mechanist what he wants, and would be compatible with an open-ended and creative view of the human mind. The hardline mechanist is offering an explanation of the 'Why Necessarily?' type, in contrast to others, including some connectionists, who merely seek to explain How Possibly the brain could work.[3] The latter may be called mechanists, but their conclusions, though illuminating, do not have determinist consequences, any more than evolutionists who explain how reptiles evolved into birds are committed to holding that the evolutionary path actually taken was one that had to go in that pre-determined way.

---

[3] I owe this crucial distinction to W.H. Dray. See his 'Explanatory narrative in history', *The Philosophical Quarterly*, 4: 15–27, 1954; or his *Laws and Explanation in History*, Oxford: Oxford University Press, 1957, ch. 6.

8

The impossibility of ultimate responsibility?

GALEN STRAWSON

1. The Basic Argument

You set off for a shop on the evening of a national holiday, intending to buy a cake with your last five-pound note to supplement the preparations you've already made. There's one cake left in the shop and it costs five pounds; everything is closing down. On the steps of the shop someone is shaking a box, collecting money for Oxfam. You stop, and it seems clear to you that it is *entirely up to you* what you do next. It seems clear to you that you are truly, radically free to choose, in such a way that you will be ultimately morally responsible for whatever you do choose.

There is, however, an argument, which I will call the Basic Argument, which appears to show that we can never be truly or ultimately morally responsible for our actions. According to the Basic Argument, it makes no difference whether determinism is true or false.

The central idea can be quickly conveyed:

(A) Nothing can be *causa sui* – nothing can be the cause of itself.
(B) In order to be truly or ultimately morally responsible for one's actions one would have to be *causa sui*, at least in certain crucial mental respects.
(C) Therefore no one can be truly or ultimately morally responsible.

We can expand the argument as follows:

(1) Interested in free action, we're particularly interested in actions performed for a reason (as opposed to reflex actions or mindlessly habitual actions).

(2)   When one acts for a reason, what one does is a function of how one is, mentally speaking. (It's also a function of one's height, one's strength, one's place and time, and so on; but it's the mental factors that are crucial when moral responsibility is in question.)

(3)   So if one is to be truly responsible for how one acts, one must be truly responsible for how one is, mentally speaking – at least in certain respects.

(4)   But to be truly responsible for how one is, in any mental respect, one must have brought it about that one is the way one is, in that respect. And it's not merely that one must have caused oneself to be the way one is, in that respect. One must also have consciously and explicitly chosen to be the way one is, in that respect, and one must have succeeded in bringing it about that one is that way.

(5)   But one can't really be said to choose, in a conscious, reasoned, fashion, to be the way one is in any respect at all, unless one already exists, mentally speaking, already equipped with some principles of choice, 'P1' – preferences, values, ideals – in the light of which one chooses how to be.

(6)   But then to be truly responsible, on account of having chosen to be the way one is, in certain mental respects, one must be truly responsible for one's having the principles of choice P1 in the light of which one chose how to be.

(7)   But for this to be so one must have chosen P1, in a reasoned, conscious, intentional fashion.

(8)   But for this to be so one must already have had some principles of choice P2, in the light of which one chose P1.

(9)   And so on. Here we are setting out on a regress that we cannot stop. True self-determination is impossible because it requires the actual completion of an infinite series of choices of principles of choice.

(10)   So true moral responsibility is impossible, because it requires true self-determination, as noted in (3).[1]

This may seem contrived, but essentially the same argument can be given in a more natural form.

---

[1] Wouldn't it be enough if one simply endorsed the way one found oneself to be, mentally, in the relevant respects, without actually changing anything? Yes, if one were ultimately responsible for having the principles in the light of which one endorsed the way one found oneself to be. But how could this be?

(1) It's undeniable that one is the way one is, initially, as a result of heredity and early experience, and it's undeniable that these are things for which one can't be held to be in any way responsible (morally or otherwise).

(2) One can't at any later stage of life hope to accede to true or ultimate moral responsibility for the way one is by trying to change the way one already is as a result of one's genetic inheritance and previous experience.

For

(3) Both the particular way in which one is moved to try to change oneself, and the degree of one's success in one's attempt to change, will be determined by how one already is as a result of one's genetic inheritance and previous experience.

And

(4) Any further changes that one can bring about only after one has brought about certain initial changes will in turn be determined, via the initial changes, by one's genetic inheritance and previous experience.

(5) This may not be the whole story, and there may be changes in the way one is that can't be traced to one's genetic inheritance and experience but rather to the influence of indeterministic factors. It is, however, absurd to suppose that indeterministic factors, for which one is obviously not responsible, can contribute in any way to one's being truly morally responsible for how one is.

2. Ultimate moral responsibility

But what is this supposed 'true' or 'ultimate' moral responsibility? An old story may be helpful. As I understand it, it's responsibility of such a kind that, if we have it, then it *makes sense* to suppose that it could be just to punish some of us with (eternal) torment in hell and reward others with (eternal) bliss in heaven. The stress on the words 'makes sense' is important, because one certainly doesn't have to believe in any version of the story of heaven and hell in order to understand, or indeed believe in, the kind of true or ultimate moral responsibility that I'm using the story to illustrate. A less colourful way to convey the point, perhaps, is to say that true or ultimate responsibility exists if punishment and reward can be fair without having any sort of pragmatic justification whatever.

One certainly doesn't have to refer to religious faith in order to describe the sorts of everyday situation that give rise to our belief in such responsibility. Choices like the one with which I began (the cake or the collection box) arise all the time, and constantly refresh our conviction about our responsibility. Even if one believes that determinism is true, in such a situation, and that one will in five minutes' time be able to look back and say that what one did was determined, this doesn't seem to undermine one's sense of the absoluteness and inescapability of one's freedom, and of one's moral responsibility for one's choice. Even if one accepts the validity of the Basic Argument, which concludes that one can't be in any way ultimately responsible for the way one is and decides, one's freedom and true moral responsibility seem, in the moment, as one stands there, obvious and absolute.

Large and small, morally significant or morally neutral, such situations of choice occur regularly in human life. I think they lie at the heart of the experience of freedom and moral responsibility. They're the fundamental source of our inability to give up belief in true or ultimate moral responsibility. We may wonder why human beings experience these situations of choice as they do; it's an interesting question whether any possible cognitively sophisticated, rational, self-conscious agent must experience situations of choice in this way.[2] But these situations of choice are the experiential rock on which the belief in ultimate moral responsibility is founded.

Most people who believe in ultimate moral responsibility take its existence for granted, and don't ever entertain the thought that one needs to be ultimately responsible for the way one *is* in order to be ultimately responsible for the way one *acts*. Some, however, reveal that they see its force. E.H. Carr states that 'normal adult human beings are morally responsible for their own personality' (Carr 1961: 89). Sartre holds that 'man is responsible for what he is' (Sartre 1946: 29) and seeks to give an account of how we 'choose ourselves' (Sartre 1943: 440, 468, 503). In a later interview he judges his earlier assertions about freedom to be incautious – 'When I read this, I said to myself: it's incredible, I actually believed that!' – but he still holds that 'in the end one is always responsible for what is made of one' (Sartre 1970: 22). Kant puts it clearly when he claims that

---

[2] See, e.g., MacKay (1960); Strawson (2010: 246–50; 1986: 281–6). When I cite a work I give the date of first publication, or occasionally the date of composition, while the page reference is to the edition listed in the bibliography.

> man *himself* must make or have made himself into whatever, in a moral sense, whether good or evil, he is to become. Either condition must be an effect of his free choice; for otherwise he could not be held responsible for it and could therefore be *morally* neither good nor evil.
>
> (Kant 1793: 40)

Since he is committed to belief in radical moral responsibility, Kant holds that such self-creation does indeed take place, and writes accordingly of 'man's character, which he himself creates' (Kant 1788: 101), and of 'knowledge of oneself as a person who . . . is his own originator' (Kant 1800: 213). John Patten, a former British Secretary of State for Education, claims that 'it is . . . self-evident that as we grow up each individual chooses whether to be good or bad' (Patten 1992).[3] Robert Kane, an eloquent recent defender of this view, writes as follows:

> if . . . a choice issues from, and can be sufficiently explained by, an agent's character and motives (together with background conditions), then to be ultimately responsible for the choice, the agent must be at least in part responsible by virtue of choices or actions voluntarily performed in the past for having the character and motives he or she now has.
>
> (Kane 2000: 317–18)

Christine Korsgaard agrees: 'judgements of responsibility don't really make sense unless people create themselves' (Korsgaard 2009: 20).

Most of us, as remarked, never actually follow this line of thought. It seems, though, that we do tend, in some vague and unexamined fashion, to think of ourselves as responsible for – answerable for – how we are. The point is somewhat delicate, for we don't ordinarily suppose that we have gone through some sort of active process of self-determination at some past time. It seems nevertheless that we do unreflectively experience ourselves, in many respects, rather as we might experience ourselves if we did believe that we had engaged in some such activity of self-determination; and we may well also think of others in this way.

Sometimes a part of one's character – a desire or tendency – may strike one as foreign or alien. But it can do this only against a background of character traits that aren't experienced as foreign, but are rather 'identified' with. (It's only relative to such a background that a character trait can stand out as alien.) Some feel tormented by impulses that they experience as alien,

---

[3] Nussbaum has something much less dramatic in mind, I think, when she writes that 'one's own character is one's own responsibility and not that of others' (Nussbaum 2004).

but in many a sense of general identification with their character pre-dominates, and this identification seems to carry within itself an implicit sense that one is generally speaking in control of, or at least answerable for, how one is (even, perhaps, for aspects of one's character that one doesn't like). So it seems that we find, semi-dormant in common thought, an implicit recognition of the idea that true or ultimate moral responsibility for one's actions (for what one does) does somehow involve responsibility for how one is: it seems that ordinary thought is ready to move this way under pressure.

There are also many aspects of our ordinary sense of ourselves as morally responsible free agents that we don't feel to be threatened in any way by the fact that we can't be ultimately responsible for how we are. We readily accept that we are products of our heredity and environment without feeling that this poses any threat to our freedom and moral responsibility at the time of action. It's natural to feel that if one is fully consciously aware of oneself as able to choose in a situation of choice, then this is already entirely sufficient for one's radical freedom of choice – whatever else is or is not the case (see further the penultimate paragraph of this chapter). It seems, then, that our ordinary conception of moral responsibility may contain mutually inconsistent elements. If this is so, it is a profoundly important fact (it would explain a great deal about the character of the philosophical debate about free will). But these other elements in our ordinary notion of moral responsibility, important as they are, are not my present subject.[4]

3. Restatement of the Basic Argument

I want now to restate the Basic Argument in very loose – as it were conversational – terms. New forms of words allow for new forms of objection, but they may be helpful nonetheless – or for that reason.

---

[4] For some discussion of the deep ways in which we're naturally compatibilist in our thinking about free will or moral responsibility, and don't feel that it is threatened either by determinism or by our inability to be self-determining, see Strawson 2010: §6.4 ('Natural compatibilism'). Clarke and Fischer are prominent among those who misrepresent my position on free will to the extent that they focus only on the line of thought set out in the current paper. See, e.g., Clarke (2005), Fischer (2006).

(1) You do what you do, in any situation in which you find yourself, because of the way you are.

So

(2) To be truly morally responsible for what you do you must be truly responsible for the way you are – at least in certain crucial mental respects.

Or:

(1) When you act, what you do is a function of how you are (what you do won't count as an action at all unless it flows appropriately from your beliefs, preferences, and so on).

Hence

(2) You have to get to have some responsibility for how you are in order to get to have some responsibility for what you intentionally do.

Once again I take the qualification about 'certain mental respects' for granted. Obviously one isn't responsible for one's sex, basic body pattern, height, and so on. But if one weren't responsible for anything about oneself, how could one be responsible for what one did, given the truth of (1)? This is the fundamental question, and it seems clear that if one is going to be responsible for any aspect of oneself, it had better be some aspect of one's mental nature.

I take it that (1) is incontrovertible, and that it is (2) that must be resisted. For if (1) and (2) are conceded the case seems lost, because the full argument runs as follows:

(1) You do what you do because of the way you are.[5]

---

[5] During the symposium on free will at the British Academy in July 2010, J.R. Lucas objected that this claim involved an equivocation. He suggested that it operated simultaneously as a conceptual claim and as a causal claim, in a way which vitiated it. I agree that it is both a conceptual claim and a causal claim, but not that this vitiates it. The following is a *conceptual* truth about the *causation* of intentional action: that with regard to the respect in which it is true to say that the action is intentional, it must be true to say that the agent does what he does – given, as always, the situation in which he finds himself or takes himself to be – because (this is a causal 'because') of the way he is; and indeed because of the way he is in some mental respect; whatever else is true, and whatever else may be going on. The truth of this claim is wholly compatible with the fact that the way you are when you act is a function of many things, including of course your experience of your situation – which is part of the way you

So

(2) To be truly morally responsible for what you do you must be truly responsible for the way you are – at least in certain crucial mental respects.

But

(3) You can't be truly responsible for the way you are, so you can't be truly responsible for what you do.

Why can't you be truly responsible for the way you are? Because

(4) To be truly responsible for the way you are, you must have intentionally brought it about that you are the way you are, and this is impossible.

Why is it impossible? Well, suppose it isn't. Suppose

(5) You have somehow intentionally brought it about that you are the way you now are, and that you have brought this about in such a way that you can now be said to be truly responsible for being the way you are now.

For this to be true

(6) You must already have had a certain nature N in the light of which you intentionally brought it about that you are as you now are.

But then

(7) For it to be true that you are truly responsible for how you now are, you must be truly responsible for having had the nature N in the light of which you intentionally brought it about that you are the way you now are.

So

(8) You must have intentionally brought it about that you had that nature N, in which case you must have existed already with a prior nature in the light of which you intentionally brought it about that you had the

___

are mentally speaking. Certainly the way you are mentally speaking isn't just a matter of your overall character or personality, and the present argument has its full force even for those who question or reject the explanatory viability of the notion of character when it comes to the explanation of action (see, e.g., Harman (1999, 2000); Doris (2002); see also Note 7 below).

nature N in the light of which you intentionally brought it about that you are the way you now are . . .

Here one is setting off on the regress again. Nothing can be *causa sui* in the required way. Even if this attribute is allowed to belong (unintelligibly) to God, it can't plausibly be supposed to be possessed by ordinary finite human beings. 'The *causa sui* is the best self-contradiction that has been conceived so far', as Nietzsche remarked in 1886:

> it is a sort of rape and perversion of logic. But the extravagant pride of man has managed to entangle itself profoundly and frightfully with just this nonsense. The desire for 'freedom of the will' in the superlative metaphysical sense, which still holds sway, unfortunately, in the minds of the half-educated; the desire to bear the entire and ultimate responsibility for one's actions oneself, and to absolve God, the world, ancestors, chance, and society involves nothing less than to be precisely this *causa sui* and, with more than Baron Münchhausen's audacity, to pull oneself up into existence by the hair, out of the swamps of nothingness . . .

> (Nietzsche 1886, §21)

The rephrased argument is essentially exactly the same as before, although the first two steps are now more simply stated. Can the Basic Argument simply be dismissed? Is it really of no importance in the discussion of free will and moral responsibility, as some have claimed? (No and No.) Shouldn't any serious defence of free will and moral responsibility thoroughly acknowledge the respect in which the Basic Argument is valid before going on to try to give its own positive account of the nature of free will and moral responsibility? Doesn't the argument go to the heart of things if the heart of the free will debate is a concern about whether we can be truly morally responsible in the absolute way that we ordinarily suppose? (Yes and Yes.)

We are what we are, and we can't be thought to have made ourselves *in such a way* that we can be held to be free in our actions *in such a way* that we can be held to be morally responsible for our actions *in such a way* that any punishment or reward for our actions is ultimately just or fair. Punishments and rewards may seem deeply appropriate or intrinsically 'fitting' to us; many of the various institutions of punishment and reward in human society appear to be practically indispensable in both their legal and non-legal forms. But if one takes the notion of justice that is central to our intellectual and cultural tradition seriously, then the consequence of the Basic Argument is that there is a fundamental sense in which no punishment or reward is ever just. It is exactly as just to punish or reward people for their actions as it is to punish or reward them for the (natural) colour of their hair or the (natural) shape of their faces. The conclusion seems intolerable, but inescapable.

Darwin develops the point as follows in a notebook entry for 6 September 1838:

> The general delusion about free will obvious ... One must view a [wicked] man like a sickly one – We cannot help loathing a diseased offensive object, so we view wickedness. – it would however be more proper to pity than to hate & be disgusted. with them. Yet it is right to punish criminals; but solely to *deter* others ... This view should teach one profound humility, one deserves no credit for anything. (yet one takes it for beauty and good temper), nor ought one to blame others. – This view will do no harm, because no one can be really *fully* convinced of its truth. except man who has thought very much, & he will know his happiness lays in doing good & being perfect, & therefore will not be tempted, from knowing every thing he does is independent of himself[,] to do harm.[6]

4. Response to the Basic Argument

I've suggested that it is step (2) of the restated Basic Argument that must be rejected, and of course it can be rejected, because the phrases 'truly responsible' and 'truly morally responsible' can be defined in many ways. I'll sketch three sorts of response.

(I) The first response is *compatibilist*. Compatibilists say that one can be a free and morally responsible agent even if determinism is true. They claim that one can correctly be said to be truly responsible for what one does, when one acts, just so long as one is in control of one's action in the way that we take an ordinary person to be in ordinary circumstances: one isn't, for example, caused to do what one does by any of a certain set of constraints (kleptomaniac impulses, obsessional neuroses, desires that are experienced as alien, post-hypnotic commands, threats, instances of *force majeure*, and so on). Compatibilists don't impose any requirement that one should be truly responsible for how one is, so step (2) of the Basic Argument comes out as false, on their view. They think one can be fully morally responsible even if the way one is is entirely determined by factors entirely outside one's control. They simply reject the Basic Argument. They

---

[6] Darwin (1838: 608). For 'wicked' in the first line Darwin has 'wrecked' (a characteristic slip).

know that the kind of responsibility ruled out by the Basic Argument is impossible, and conclude that it can't be the kind of responsibility that is really in question in human life, because (they insist) we are indeed genuinely morally responsible agents. No theory that concludes otherwise can possibly be right, on their view.

(II) The second response is *libertarian*. *Incompatibilists* believe that freedom and moral responsibility are incompatible with determinism, and some incompatibilists are libertarians, who believe that we are free and morally responsible agents, and that determinism is therefore false. Robert Kane, for example, allows that we may act responsibility from a will already formed, but argues that the will must in this case be

> 'our own' free will by virtue of other past 'self-forming' choices or other actions that were undetermined and by which we made ourselves into the kinds of persons we are . . . [T]hese undetermined self-forming actions (SFAs) occur at those difficult times of life when we are torn between competing visions of what we should do or become.
>
> (Kane 2000: 318–19)

They paradigmatically involve a conflict between moral duty and non-moral desire, and it is essential that they involve indeterminism, on Kane's view, for this 'screens off complete determination by influences of the past' (Kane 2000: 319). He proposes that we are in such cases of 'moral, prudential and practical struggle . . . truly "making ourselves" in such a way that we are ultimately responsible for the outcome', and that this 'making of ourselves' means that 'we can be ultimately responsible for our present motives and character by virtue of past choices which helped to form them and for which we were ultimately responsible' (Kane 1989: 252).

Kane, then, accepts step (2) of the Basic Argument, and challenges step (3) instead. He accepts that we have to 'make ourselves', and so be ultimately responsible for ourselves, in order to be morally responsible for what we do; and he thinks that this requires indeterminism. But the old, general objection to libertarianism recurs. How can indeterminism possibly help with moral responsibility? How can the occurrence of indeterministic or partly random events contribute to my being truly or ultimately morally responsible either for my actions or for my character? If my efforts of will shape my character in a positive way, and are in so doing essentially partly indeterministic in nature, while also being shaped (as Kane grants) by my already existing character, why am I not merely *lucky*?

This seems to be a very strong general objection to any libertarian account of free will. Suppose, in the light of this, that we put aside the Basic

Argument for a moment, and take it as given that there is – that there must be – some respectable sense in which human beings are or can be genuinely morally responsible for their actions. If we then ask what sort of account of moral responsibility this will be, compatibilist or incompatibilist, I think we can safely reply that it will have to be compatibilist. This is because it seems so clear that nothing can ever be gained, in an attempt to defend moral responsibility, by assuming that determinism is false.

(III) The third response begins by accepting that one can't be held to be ultimately responsible for one's character or personality or motivational structure. It accepts that this is so whether determinism is true or false. It then directly challenges step (2) of the Basic Argument. It appeals to a certain picture of the *self* in order to argue that one can be truly free and morally responsible in spite of the fact that one can't be held to be ultimately responsible for one's character or personality or motivational structure.

It can be set out as follows. One is free and truly morally responsible because one's self is, in a crucial sense, independent of one's character or personality or motivational structure – one's CPM, for short. Suppose one is in a situation which one experiences as a difficult choice between A, doing one's duty, and B, following one's non-moral desires. Given one's CPM, one responds in a certain way. One's desires and beliefs develop and interact and constitute reasons in favour both of A and of B, and one's CPM makes one tend towards either A or B. So far, the problem is the same as ever: whatever one does, one will do what one does because of the way one's CPM is, and since one neither is nor can be ultimately responsible for the way one's CPM is, one can't be ultimately responsible for what one does.

Enter one's self, S. S is imagined to be in some way independent of one's CPM. S (i.e. one) considers the deliverances of one's CPM and decides in the light of them, but it – S – incorporates a power of decision that is independent of one's CPM in such a way that one can after all count as truly and ultimately morally responsible in one's decisions and actions, even though one isn't ultimately responsible for one's CPM. The idea is that step (2) of the Basic Argument is false because of the existence of S (for a development of this view see, for example, Campbell 1957).

The trouble with the picture is obvious. S (i.e. one) decides on the basis of the deliverances of one's CPM. But whatever S decides, it decides as it does because of the way it is (or because of the occurrence in the decision process of indeterministic factors for which it – i.e. one – can't be responsible, and which can't plausibly be thought to contribute to its ultimate moral responsibility). And this brings us back to where we started. To be a source of ultimate responsibility, S must be responsible for being the way it is. But this is impossible, for the reasons given in the Basic Argument. So while the story of S and CPM adds another layer to the description of the human decision process, it can't change the fact that human beings cannot be ultimately self-determining in such a way as to be ultimately morally responsible for how they are, and thus for how they decide and act.

In spite of all these difficulties, many of us (nearly all of us) continue to believe that we are truly morally responsible agents in the strongest possible sense. Many of us, for example, feel that our capacity for fully explicit self-conscious deliberation in a situation of choice suffices – all by itself – to constitute us as such. All that is needed for true or ultimate responsibility, on this view, is that one is in the moment of action *fully self-consciously aware of oneself as an agent facing choices*. The idea is that such full self-conscious awareness somehow renders irrelevant the fact that one neither is nor can be ultimately responsible for any aspect of one's mental nature: the mere fact of one's self-conscious presence in the situation of choice can confer true moral responsibility. It may be undeniable that one is, in the final analysis, wholly constituted as the sort of person one is by factors for which one cannot be in any way ultimately responsible, but the threat that this fact appears to pose to one's claim to true moral responsibility is, on this view, simply annihilated by one's self-conscious awareness of one's situation.

This is an extremely natural intuition; but the Basic Argument appears to show that it is a mistake. For however self-consciously aware we are as we deliberate and reason, every act and operation of our mind happens as it does as a result of features for which we are ultimately in no way responsible. And yet the conviction that self-conscious awareness of one's situation can be a sufficient foundation of strong free will is extremely powerful. It runs deeper than rational argument, and it survives untouched, in the everyday conduct of life, even after the validity of the Basic Argument has been admitted. Nor, probably, should we wish it otherwise.[7]

---

[7] On this last point, see, e.g., P.F. Strawson (1962); for a doubt, see Smilansky (1994). It will be interesting to see how the conviction of free will stands up to increasing public awareness of results in experimental and social psychology, which show that our actions are often strongly influenced by factors, situational or otherwise, of which we are completely unaware (see, e.g.,

References

Campbell, C.A. (1957) Has the self 'free will'? In C.A. Campbell, *On Selfhood and Godhood*, London: Allen & Unwin.

Carr, E.H. (1961) *What Is History?*, London: Macmillan.

Clarke, R. (2005) On an argument for the impossibility of moral responsibility, *Midwest Studies in Philosophy*, 19: 13–24.

Darwin, C. (1987 [1838]) *Charles Darwin's Notebooks, 1836–44*, Cambridge: Cambridge University Press.

Doris, J. (2002) *Lack of Character: Personality and Moral Behavior*, Cambridge: Cambridge University Press.

Fischer, J. (2006) The cards that are dealt you, *The Journal of Ethics*, 10: 107–29.

Harman, G. (1999) Moral philosophy meets social psychology: virtue ethics and the fundamental attribution error, *Proceedings of the Aristotelian Society*, 99: 315–31.

Harman, G. (2000) The nonexistence of character traits, *Proceedings of the Aristotelian Society*, 100: 223–6.

Kane, R. (1989) Two kinds of incompatibilism, *Philosophy and Phenomenological Research*, 50: 219–54.

Kane, R. (2000) Free will and responsibility: ancient dispute, new themes, *The Journal of Ethics*, 4: 315–22.

Kant, I. (1956 [1788]) *Critique of Practical Reason*, trans. L.W. Beck, Indianapolis: Bobbs-Merrill.

Kant, I. (1960 [1793]) *Religion within the Limits of Reason Alone*, trans. T.M. Greene and H.H. Hudson, New York: Harper & Row.

Kant, I. (1993 [1800]) *Opus postumum*, trans. E. Förster and M. Rosen, Cambridge: Cambridge University Press.

Knobe, J. and Nichols, S. (2008), *Experimental Philosophy*, New York: Oxford University Press.

Korsgaard, C. (2009) *Self-Constitution: Agency, Identity, and Integrity*, Oxford: Oxford University Press.

Doris (2002), Wilson (2002), Nahmias (2007), Knobe and Nichols (2008). The general effect of this 'situationist' line of enquiry is to cast increasing doubt on our everyday picture of ordinary adult human agents as consciously aware of, and in control of, themselves and their motivations and subsequent actions in such a way that they are, generally speaking, fully morally responsible for what they do. Situationism finds a natural ally in Freudian theory, while considerably extending the range of factors that threaten to undermine our everyday picture of responsibility. It tells us that we are far more 'puppets of circumstances' than we realize; it questions our conception of ordinary human beings as genuinely free agents in a way that is independent of any considerations about determinism or the impossibility of self-origination. At the same time (again in line with Freudian theory) it grounds a sense in which greater self-knowledge, a better understanding of what motivates one, can increase one's control of and responsibility for one's actions.

MacKay, D.M. (1960) On the logical indeterminacy of free choice, *Mind*, 69: 31–40.

Nahmias, E. (2007) Autonomous agency and social psychology. In M. Marraffa *et al.* (eds) *Cartographies of the Mind: Philosophy and Psychology in Intersection*, Dordrecht: Springer.

Nietzsche, F. (1966 [1886]) *Beyond Good and Evil*, trans. Walter Kaufmann, New York: Random House.

Nussbaum, M. (2004) Discussing disgust: on the folly of gross-out public policy: an interview with Martha Nussbaum, *Reason*, 15 July 2004.

Patten, J. (1992) article in *The Spectator*, 16 April 1992.

Sartre, J.-P. (1969 [1943]) *Being and Nothingness*, trans. Hazel E. Barnes, London: Methuen.

Sartre, J.-P. (1970) interview in *New Left Review*, 58, reprinted in *New York Review of Books*, 26 March 1970, p. 22.

Sartre, J.-P. (1989 [1946]) *Existentialism and Humanism*, trans. Philip Mairet, London: Methuen.

Smilansky, S. (1994) The ethical advantages of hard determinism, *Philosophy and Phenomenological Research*, 54: 355–63.

Strawson, G. (1986) *Freedom and Belief*, Oxford: Clarendon.

Strawson, G. (2010) *Freedom and Belief*, Second edition, Oxford: Clarendon.

Strawson, P.F. (1974 [1962]) Freedom and resentment. In P.F. Strawson, *Freedom and Resentment*, London: Methuen.

Wilson, T. (2002) *Strangers to Ourselves: Discovering the Adaptive Unconscious*, Cambridge, MA: Harvard University Press.

9
Moral responsibility and the concept of agency

HELEN STEWARD

> It is strange that philosophers have been able to argue endlessly about determinism and free-will, to cite examples in favour of one or the other thesis without ever attempting first to make explicit the structures contained in the very idea of action.
>
> (Sartre 1958: 433)

There is a long tradition of supposing that moral responsibility must be inconsistent with a deterministic view of the universe. In this chapter, I shall follow in that long libertarian tradition, and insist, as many others have done, on the incompatibility of moral responsibility and determinism. But if libertarianism is not to be too easily vanquished, I think it is crucially important to be clear about exactly where the source of the incompatibility really lies. Traditionally, the argument for incompatibility has gone by way of the principle that an agent cannot be morally responsible for what she has done unless she could have done otherwise, incompatibilists believing that determinism would be inconsistent with agents possessing this essential power of having done something different. But in recent years, this principle – usually dubbed the Principle of Alternate Possibilities, or PAP – has fallen into widespread disfavour. Harry Frankfurt, in particular, is supposed by many to have shown conclusively, by means of an ingenious counterexample, that the principle is simply not true (Frankfurt 1969). In this paper, though, I shall try to explain why I think the real lesson of Frankfurt's example lies elsewhere. What it shows us is not that moral responsibility is consistent with determinism after all. What it shows is rather that the power to do otherwise which moral responsibility depends upon is not quite the power it has usually been taken to be, and moreover is important for a reason quite different from the one which is usually emphasized. The reason usually emphasized is a moral reason – that it is unfair to hold an agent morally responsible for what she cannot help.[1] But the real reason why determinism and moral responsibility are inconsistent is, I shall argue, not moral, but metaphysical. The real reason is that determinism is inconsistent with *agency*, which is a necessary (though not, of course, a sufficient) condition of moral responsibility.

1. Frankfurt and the Principle of Alternate Possibilities

Let me begin with a quick reminder of Frankfurt's counterexample. We are supposed to imagine an agent, Jones, who is considering doing something – say, voting Labour in the forthcoming election. Black, a second agent, would very much like Jones to vote Labour in the election – and, Frankfurt tells us, 'is prepared to go to considerable lengths to get his way' (Frankfurt 1969: 835). In what has become the most popular version of the case, we are to imagine that Black is an exceedingly clever neurosurgeon who has implanted a device in Jones's brain by which he can monitor Jones's thought processes, and by means of which (we are told) he would also be able to ensure, if necessary, that Jones will decide to vote, and then will vote, as he, Black, wishes.[2] However, Black would also greatly prefer *not* to intervene if there is any hope of avoiding it. He waits, therefore, until Jones is about to make up his mind what to do, scanning Jones's brainwaves by means of his clever device, for signs of the impending decision. If there is any sign that Jones is about to form an intention to vote for anyone else, or to form an intention not to vote at all, Black will intervene at that moment to ensure that Jones instead forms the intention to vote Labour – and will then continue to monitor Jones's thought processes to ensure that he actually follows through in the polling booth. But in fact, in the event, there is no such sign. Jones simply makes up his mind to vote Labour, and then does so unhesitatingly, without Black having had to

---

[1] Though people's intuitions that this is so are usually much stronger in the case of blame than they are in the case of praise – one sign, perhaps, that all is not quite right with the traditional line of argument.

[2] Later I shall question whether this description of the extent of Black's power over Jones is really coherent.

intervene at all. In this circumstance, Frankfurt argues, we would surely regard Jones as morally responsible for having voted the way he did. After all, in the event, Black did nothing whatever. But it is also true, according to Frankfurt, that Jones could not have done other than vote Labour. For if he had shown any inclination to do anything other than vote Labour, Black would have intervened, and Jones would have ended up voting Labour in any case. The Principle of Alternate Possibilities, then, which says that an agent can be morally responsible for what she has done only if she could have done otherwise, must be false. The power to do otherwise cannot be a necessary condition of moral responsibility.

I will not attempt to offer any kind of comprehensive overview of the vast number of different responses that this purported counterexample to PAP has generated. The provision of such an overview would be too large a task to be confined within the limits of a volume such as this. I want instead to concentrate on developing what I take to be the moral of Frankfurt's story; and though I think my view has certain things in common with various responses which have been made before to Frankfurt's example, it is not quite the same as any of them. Moreover the view I shall outline brings to the surface an important intuition which I think is present, but has often been left submerged in the views of others. I should perhaps say that what I shall offer here is the briefest of brief overviews of a position on free will each of whose parts really needs much more in the way of defence than I am able to provide here. But I hope I shall be able to say enough, at any rate, to make it persuasive that a position with the shape I shall characterize is possible – and that it provides a path worthy of further exploration for the libertarian.

Let us begin, then, by asking why anyone would ever have thought that moral responsibility should require the ability to do otherwise. Here is what I think is a very common line of thought: it would be simply *unfair* to hold anyone responsible for what they cannot help, for doing something they could not have avoided doing. 'I couldn't help it', 'there was no alternative', 'I was forced', etc. are common excuses – and a generalization from these practices of exculpation might make us think that the general principle that a person cannot be held responsible for what they cannot help, for reasons of fairness, was highly plausible. But Frankfurt's argument casts considerable doubt on this line of thinking, as Frankfurt himself was at pains to emphasize in his original article. Jones had (unbeknownst to him) no alternative to voting Labour – there is a sense, then, in which it would be right to say that he could not help doing so – or at any rate, that he could not avoid it.[3] But it does not follow, it seems, that it would not be fair to hold him morally responsible for what he did. For in fact, nothing interfered with his normal processes of deliberation and decision-making. Jones's lack of alternatives in this case seems to have nothing whatever to do with the question whether or not he is responsible for what he does. If the Principle of Alternative Possibilities is to be thought of as based upon these ideas concerning fairness, then, it seems we must reject it. Frankfurt himself claims that 'The doctrine that coercion excludes moral responsibility is not correctly understood . . . as a particularized version of the principle of alternate possibilities'.[4] That is right – but for my purposes, it is better to put things the other way around. The Principle of Alternate Possibilities is not correctly understood as a generalized version of our principles concerning the unfairness of blame in cases of coercion, psychological compulsion, etc.

How, then, is it to be understood? The suggestion of many writing in the wake of Frankfurt's work, of course, is that it is to be understood simply as a mistaken *over*-generalization from plausible principles of exculpation – an over-generalization which we ought now to reject, having seen that it cannot be maintained – and that with its rejection, compatibilism about moral responsibility and determinism is set fair to triumph, since the condition with which determinism was judged by some to be inconsistent has turned out not to be a necessary condition of moral responsibility after all. But I want here to make a rather different suggestion. In my view, Frankfurt's example merely highlights rather neatly that the Principle of Alternate Possibilities as usually formulated (i.e. as the principle that an agent S is morally responsible for what she has done only if she could have done otherwise) does not properly and unambiguously capture the

---

[3] In fact, I think, the much greater naturalness of the latter over the former locution may tell us something. It would be odd, I suggest, to say of Jones, who went ahead and voted Labour in a perfectly ordinary way that he 'couldn't help it'. 'He couldn't help it' is most naturally understood to be a comment on a person's actual action – not a remark concerning their overall modal situation – and thus understood, it seems inappropriate to use it of Jones, whose actual action was (we may suppose for the sake of argument) fully deliberate and entirely voluntary. What this fact usefully reveals is that – *pace* Frankfurt – we sometimes want to speak of a person's powers in relation to something *particular* – the event which is their action or movement itself – not merely in respect of something general – their act considered as a *type*. Jones could not have avoided voting Labour (the 'act type'). But *this* act of voting (it) was arguably something he could have helped. This distinction between types of action and particular doings is a distinction which my view exploits – as will shortly be seen.
[4] Frankfurt (1969: 831).

premise from which the best argument for libertarianism ought to begin. I suggest that in order to arrive at a preferable formulation, rather than beginning with a thought about what fairness demands, we should begin the libertarian argument with a thought about what *agency* requires.

2. Agency and determinism

To understand how this thought might arise, and to see the intuitions from which it flows, consider for a moment the world as it is supposed to be according to the determinist, a series of events and states of affairs, inexorably superseded by other events and states of affairs, according to ineluctable laws of nature. I suggest that one might well feel that this kind of world is not correctly characterized as a world in which there are agents doing things they could not have avoided doing (as traditional versions of libertarianism tend to suggest). I suggest that this world is best characterized as a world in which there are *no agents at all* – in which there is simply no space whatever for the entities we think of as agents, entities that things can be *up to*, entities which can hold portions of the fate of the universe in their hands.[5] How can anything be up to an entity in a world where

[5] It has been suggested to me (by an anonymous reviewer of this paper) that perhaps the claim that an agent must be the sort of thing that various matters can be 'up to' already begs the question against the compatibilist – and that it would be less tendentious merely to say that an agent is something that can act. But to say that an agent is something that can act does nothing other than to connect a noun to an associated verb, and merely raises the question which things may be said to *act*. Can a wave act? Or a computer? If not, why not? It might perhaps be said that actions are events which are caused by appropriate types of mental event or state – things like decisions and intentions. But this merely pushes the question back. We now need to know to which things we are genuinely entitled to take what Dennett (1971, 1987) has called the intentional stance – to which things we may apply various of the psychological concepts by means of which we organize our understanding of certain of the entities we come across in the world. I should want to claim that the answer to this question is that we are truly entitled to take the stance – and to construe it realistically – towards all and only entities which are such that things can be up to them, because this sort of representational psychology is only truly essential to the explanation of a thing's activity if things may conceivably be settled *by* that entity in the light of what it thinks and wants – and thus that we have come full circle to the principle that agents are entities that certain matters can be up to.

These claims about the connections that exist between folk psychology, agency and the power to settle things (to have things be 'up to' one) are controversial, of course, and I cannot properly defend them here against all comers. But the charge specifically of begging the question against the compatibilist must be refuted. The best way to refute it, I believe, is to

everything is settled by initial conditions and the laws of nature? The compatibilist will attempt to insist that the notion of something's being 'up to' an agent can be reduced to the idea of that thing's being the causal consequence of an (appropriately formed) decision on the part of that agent – one which flows in the right sort of way, say, from her beliefs and desires. But the 'decision' itself, of course, (and the preceding beliefs and desires) are themselves simply further inevitable consequences of initial conditions and laws of nature, which will make someone who is persuaded by the sort of intuitions I am attempting to elicit ask by what right we are to call these things 'decisions' in the first place. What has the *decision-maker* to do with anything, given this picture? What is she able to settle at the time of decision, given that everything is settled already by the initial conditions and the laws? The agent herself as a genuine source of settling seems to vanish from the picture of causality that we are offered, turned into a mere place or vessel where events bring others about.[6] And for this reason, one might well think that in a deterministic world there can be no actions and there can be no agents – and so a fortiori, that there could be nothing that was morally responsible for anything. Moral responsibility is reserved to agents – and so a world which excludes agency is also a world which excludes moral responsibility. In the rest of this chapter, I want to defend this line of thinking – by showing how it is perfectly consistent with our intuitions in the Frankfurt case, by developing a little the conception of

argue that even if it is thought that the conception of agency which is in play here is implausibly rich, there can be nothing implausible about the idea that various matters have to be up to beings which possess those powers constitutive of a much more traditional candidate for compatibilist rescue – *free will*. If compatibilism is to be worthy of its name, then what it finds to be compatible with determinism must be something worth having. Free will of a sort which is consistent with things *not* being up to me seems scarcely worth wanting – it certainly seems most unlikely that any such power could underpin moral responsibility. The issue, then, is not (as the charge of question-begging suggests) whether the compatibilist should want her compatibilism to validate the claim that (certain) things are up to us; clearly, she should want this. The issue is whether the compatibilist *can* validate the claim. Any compatibilist worth her salt, I believe, will think she can do so. I shall be claiming below, of course, that she cannot – but *this* is the issue and no question has been begged against compatibilism by the assumption that 'up-to-usness' is part of what it needs to save.

[6] It might be said that it is not the determinism as such that makes for the difficulty here – and in a sense that is right, since certain sorts of *in*deterministic picture would certainly not help to solve the problem. But as I see it, there are varieties of indeterministic picture that *do* solve it; but no varieties of deterministic picture that do. When it is said that indeterminism is of no help to those in search of free will, that is because only the unhelpful pictures (involving, e.g., mere microphysical randomness) are in view.

agency in question, and finally, by answering what I expect will be the main line of objection to the claim – namely that since we do not know for sure whether or not determinism is true, my view must raise absurd doubts about the very existence of agents.

Let us begin then, by returning to Frankfurt's own example, armed with this new thought about why we might think that moral responsibility is inconsistent with determinism. Why are we so ready to concede that Jones is morally responsible for what he does? Because it is obvious that his actual decision to vote Labour is an exercise of agency.[7] Nothing is lacking to that decision and its mode of formation that would not be a feature of any perfectly ordinary and utterly unfettered instance of decision-making. And since there are no further special responsibility-undermining features in this case, if we are ever morally responsible for making decisions, Jones is responsible for this one. And equally, in the counterfactual situation in which Black is forced to intervene, it is obvious that Jones would not have been responsible for his 'decision'. Why not? Well, because this so-called decision would not have been, in that case, an exercise of agency on his part. It would have been the result of an exercise of agency, rather, on *Black's* part, a fact which makes it, I suggest, problematic to speak, in this case, of Jones's having made a decision at all, even one which was some-how brought about by Black. A decision is a variety of mental action – one which normally consists in putting oneself into a state of intention. But Jones did not put himself into a state of intention in the counterfactual scenario. Black put him (Jones) into a state of intention. It seems to me a mistake, therefore, to say that *Jones* decided anything under these circum-stances. And if this is a mistake, that has a bearing on what we ought to say about the *actual* situation. If the counterfactual situation is not one in which Jones decides to vote Labour, then we need to re-evaluate the sug-gestion that he could not have done other than decide to do this. Had he not thus decided, what would have happened would not have been that Black would have made him decide. What would have happened would have been that Black would have put him into the state of intention that

---

[7] Of course, we can be morally responsible for all sorts of things which are not *themselves* exercises of agency – such as beliefs, states of affairs, character traits, etc. But I think we can only coherently be held responsible for these other things if we can be held to have powers of agency which relate to them – for instance, that I can *examine* my beliefs and subject them to scrutiny; I can *alter* certain states of affairs; I can *transform* my character in various respects. Where such powers as these are imagined to be entirely lacking, it does not seem plausible any longer to hold an agent responsible for the things in question.

decisions normally effect. But in that case, surely Jones could have refrained from deciding to vote Labour simply by not thus deciding and thereby triggering an intervention by Black. And if he could have thus refrained, then the Frankfurt example is not a counterexample to PAP after all.

This particular way of dealing with alleged Frankfurtian counter-examples to the Principle of Alternate Possibilities will not work quite so neatly as this, unfortunately, in every kind of case. The strategy I have adopted here is dependent on the thing which Jones cannot do other than being a decision. For it is quite plausible that decisions are *essentially* actions (one cannot decide involuntarily or accidentally, for instance)[8] – and hence that nothing can count as A's decision which does not also count as A's action. This is what makes it possible to insist that the counterfactual scenario is not in fact one in which Jones decides anything at all – and hence, in turn, that the actual situation is not one in which Jones could not have refrained from thus deciding. But many Frankfurt cases relate not to such mental acts as decisions but rather to overt actions like shooting a President, say. And one cannot maintain the principle that one is morally responsible for shooting a President, only if one could have done otherwise in the face of a Frankfurt-style example – for there is no reason to deny that Black *can* make Jones shoot the President. It is not possible to insist that all shootings are essentially actions – for shootings may perfectly well be accidental or involuntary. But though Black can make Jones shoot the President, there remains, I insist, something that Black cannot do. What he cannot do is make Jones shoot the President by directly producing an event that nevertheless still counts as an exercise of agency – as an action – on Jones's part. To coin a bit of terminology – he cannot make Jones shoot$_A$ the President – where to shoot$_A$ is to shoot in such a way that that shooting is *one's own action*.[9]

---

[8] Matters pertaining to the *content* of the decision can be accidental, of course. For instance, I could decide accidentally to shoot my mother, by deciding to shoot the person who has just knocked at my door (not realizing that the person who has just knocked at my door is my mother). But the occurrence of the decision itself cannot be something that is accidental. That I have decided (something) cannot be an accident. That I have shot someone can be.

[9] Might it be said that the alternate possibility which I have alleged remains available to Jones – the power not to *act* in the way in which he did in fact act – is insufficiently 'robust' to do the work required of it here? (See Fischer (1994) for the original 'robustness' objection.) Full details of my response to this challenge can be found in Steward (2009), but in brief, my claims there are as follows. A number of quite separate demands on the wielder of PAP have been made in the name of 'robustness', some more readily justifiable than others, and it is essential to be clear about which of these demands is truly legitimate. I recognize two valid concerns:

Why not? On a certain conception of what actions *are*, this might seem puzzling. Actions, it might be said, are just physical events of certain sorts – either bodily movements, or perhaps prior neural events, or perhaps composites of the two. But whatever sort of physical event we decide actions are, there seems no reason to suppose that Black cannot organize for the occurrence of a physical event of that sort. So why can't Black's intervention make it the case that Jones has acted? The answer, I think, is that even if actions are physical events – as I agree, in some sense, they must be – their categorization as *actions on the part of a certain agent* depends not on this physical characterization, but on the relation borne by that agent to the event in question – actions are events that only exist in virtue of the fact that their *agent* has been the source of some input into the world. Actions are the beginnings of causal chains which are initiated by their agents – and so where the initiation of such a chain lies elsewhere than with a given agent, what has occurred cannot be an action *of* that agent. There are therefore conceptual conditions on the classification of any event as an action which are flouted in the case we are asked to imagine by Frankfurt. If an action is essentially an input into the course of nature *by its agent*, then the agent must possess and retain certain capacities in respect of the processes that constitute it – in particular, the bodily systems in question must be ones which are under her control to exert or not exert at the time of action – otherwise, it is not up to her whether or not the action occurs, or how, precisely, it does so; and hence she settles nothing at the time of action. There is a sense, therefore, in which we simply cannot be *directly*

(i) whatever Principle of Alternate Possibilities is endorsed by the libertarian, it should be a principle which there is at least some *prima facie* temptation to endorse, whether one is a libertarian or not; and (ii) whatever alternate possibility is alleged to exist in a Frankfurt case, it must be an alternate possibility which it is within the power of the agent to bring about, not merely an alternative outcome which might have happened. Both these concerns are met by my view. The principle on which I rely below is one which says that an agent S is morally responsible for φ-ing only if her actual φ-ing has some detailed and specific description as a D-ing, say, such that the agent could have refrained from D-ing (because only thus could her φ-ing have been an action in the first place). I believe (and argue in the chapter) that this admittedly unwieldy formulation relies in fact on a very intuitive conception of action which compatibilists as well as libertarians might have reason to want to endorse; a conception which insists that the power to act involves the power to refrain. There is evidence in the work of such classical compatibilists as Hobbes and Hume of an acceptance of this conception of action; thus concern (i) is met. And in the case described above, Jones genuinely has it within his power to refrain from *acting* in the way he does – which demonstrates that concern (ii) is also met.

manipulated to act, because the manipulation itself contravenes the conditions of agency. Where action is concerned, manipulation is of course possible – but it has to go by way of our motivational systems – we have to be persuaded, cajoled, bribed, guilt-tripped, made offers we can't refuse, etc. But my actions, I suggest, simply cannot be *directly* brought about by others. For no event that is directly brought about by another's action could be an action *of mine*.[10] Though it is not right, then, to uphold the Principle of Alternate Possibilities in its original form – i.e. that an agent is morally responsible for what she has done only if she could have done otherwise – we ought to be able to uphold some version of the following thought, even in the teeth of Frankfurt-style examples – that the existence of an action is always dependent upon the simultaneous possession by the agent of a certain power of refrainment in respect of that action – since nothing in respect of which I do not possess this power of refrainment could count as my action in the first place.

In a moment, I shall move on to say something about this power of refrainment and how it ought to be specified. Before that, though, it will be necessary to address, albeit briefly, a worry many will feel about what I have just said about the concept of an action. I have characterized actions as the beginnings of causal chains which are initiated by their agents. But it will be said that apart from random occurrences, such as, for example, the individual emissions involved in radioactive decay, which surely cannot serve as a model for actions, there are no such things as the beginnings of causal chains – that everything that ever happens in our universe can be traced back to antecedents which produce and explain it. In the next section, I shall question what reason there is to hold this view.

3. Causality, determinism and the 'beginnings' of chains

There will not be space here to say everything I should like to say about the view that there are no such things as the 'beginnings' of causal chains at the macroscopic level, much less adequately to defend what I should like to say.[11] But since it is, I suspect, likely to be the source of much scepticism about the type of view I want to defend, I must say *something*. Suffice it to

---

[10] See Alvarez (2009) for a similar view.
[11] I have made a start on the defence elsewhere – see Steward (2008).

say, then, that I believe the widespread conviction that (whatever may be the case in the realms of microphysics) at any rate *macrophysical* determinism must be true, is an outdated hangover from Newtonian visions of the universe that we philosophers mostly still remain entranced by, because we have not properly escaped the worldviews encouraged by our school-level maths and physics lessons; and I do not think that we shall ever have an acceptable metaphysics of action until we give it up. Some substances, I should like to insist – the higher animals – have powers to make certain things happen – in particular, changes to the distribution and arrangement of their own bodily parts – in ways not merely dictated by the past and the laws. Perhaps this amounts to a denial of Galen Strawson's claim that nothing can be *causa sui*;[12] though I confess I am not sure about that, because I am not exactly sure what that claim is supposed to imply. That nothing can be the cause of itself seems a reasonable enough claim – but what Strawson's argument seems to need is the stronger claim that nothing can be the cause of that same thing's coming to have a certain property – which seems a much more dubious proposition. At any rate, the idea that animals can make changes happen in the parts of their own bodies is, I think, an ancient and utterly natural view which I believe we have been encouraged to give up by a range of principles which are thought to be in conflict with it: some physical; some metaphysical; some false, like the claim that causal relations are always relations between events; some whose consequences are merely poorly understood, like the conservation laws. None of these principles, though, I believe, is truly justificatory of the idea that there cannot be what I have elsewhere called 'fresh starts'[13] – places in the structure of time and space where things occur which are not wholly attributable to what has gone before.[14] The general idea that perhaps there are some *microphysical* fresh starts is generally accepted by philosophers, inclined to believe that physics requires them because they have been told so. But what philosophers seem to find it very difficult to permit is the idea that there might be such top-down fresh starts as it is pre-theoretically natural to suppose animal actions might be – macroscopic events whereby macroscopic individuals exercise powers to make things happen in their

---

[12] See Chapter 8, this volume.

[13] Following Ross (1924: lxxxi).

[14] The word 'wholly' is important here. Causal *influence* is perfectly consistent with the existence of what I am here calling 'fresh starts'. What is necessary for a fresh start is only that what happens should not be *wholly* settled by what has gone before.

own parts. We are used to supposing that the powers of large things are merely epiphenomenal products of the powers of their small components – as seems broadly to be the case for the mechanical systems we know how to produce. But the idea that this is so is so manifestly at odds, I believe, with the structure of the phenomenon of action that we need to seek a non-mechanistic, though not, for all that, *non-naturalistic* understanding of what biology has managed to accomplish by means of the evolution of animals. An unprejudiced assessment of the facts, it seems to me, suggests that evolution has produced some entities which are able themselves to be the source of outcomes that are not fully determined in all their details by what has gone before, but which are up to the animal to settle at the time of action. I do not say it is easy to come by the emergentist-style metaphysics which might enable us to make sense of this idea. But I do say that it is far more reasonable to suppose that such a metaphysics must be available for the formulating, than to deny the things that would have to be denied were it to be true that the exercise of our agential powers were only epiphenom-enal by-products of the hum and buzz of chemical processes taking place amongst our neurons and synapses.

4. The relevant power of refrainment

Let me return, now, though, to the business I left off above, which is the elaboration of my assertion that a certain power of refrainment attends all genuine actions. I think it needs to be admitted straightaway that it is not a simple matter to say what this relevant power of refrainment *is* – what it is that the agent needs to be able to refrain *from*, as it were, if she is gen-uinely to act. The simple and tempting idea that in the case of a φ-ing which is an action – i.e. a φ_A-ing – the agent must have been able to refrain *from* φ_A-ing, is clearly wrong, I believe. Counterexamples seem to me to be provided by instances of so-called 'volitional necessity' (Frankfurt 1982: 86). Consider, for instance, a case in which I know that my children are inside a burning house and that there is no hope of their being rescued unless I go in to save them. It might seem right to say that under these circumstances, I am simply not able not to do so – I cannot refrain. But surely, if I do rush in to rescue them, then that is an action of mine – and one, moreover, for which I would clearly be morally responsible.

It might be retorted at this point that the capacity to do otherwise that is relevant to moral responsibility cannot be the kind of capacity that can be rendered ineffectual by the mere strength of an opposing motivation. Thus, for example, one might be inclined to insist that I could (in the relevant sense) have done other than have run into the burning house, even though it might be admitted that there was no realistic possibility whatever, my motivations being what they were, that anything else should have happened. This tends to be what compatibilists sympathetic to some form of the Principle of Alternate Possibilities are wont to say – that the capacity to do otherwise that is relevant to moral responsibility does not depend on the bare possibility that another thing should have happened in the precise circumstances in question, but rather on the agent's abilities thought of in some rather more general way. And thought of in this more general way, it might be said, I continue to possess those capacities – for instance, it might be insisted that I could have stayed put instead of running into the house, on the grounds that there was no one forcing my limbs to move, no external force pulling me along, etc. But though I have some sympathy with this line of thinking, ultimately I think it will not do. It is a rejoinder that underestimates the threat that determinism poses to the very idea of agency – that underestimates the power of the idea that as agents, open possibilities must exist for us at the very moment of action, t – and throughout any period t_1 to t_n for which the action persists. Compatibilists are apt to scoff at the suggestion that we might require for freedom the capacity to do something other than what we in fact go on to do, *even given* the set of motivations, reasons, emotions, etc. which attend us at the moment just prior to our action – how could it be important or valuable to us, they ask, to have the capacity to go on to do something that might, in the light of those motivations, reasons, emotions, etc., appear simply insane? It is a powerful point, and I think it ought to be conceded that the libertarian should not insist on this requirement in the form in which it is usually offered. Given alternative courses of action A and B and a set of reasons and motivations clearly favouring A, that is, we ought not to insist that there need be any possibility at the time of action that the free agent undertake course of action B instead. But it does not follow from this concession that no possibilities at all need exist at t of the sort that might make trouble for the compatibilist. A bit of terminology will prove useful here to formulate the claim I want to make. Let us say that an event or state of affairs whose occurrence or obtaining at a given time t is necessitated by certain events and states of affairs prior to t together with the laws of nature is 'historically inevitable'. The claim I want to make is that it is impossible that an event that was historically inevitable could be an action.

Note that the proposition I want to defend is not simply the claim that *facts about what we will do* cannot be historically inevitable, given our motivations and other circumstances. What I have already said implies that such facts may indeed be historically inevitable. It might, for example, be historically inevitable that I will run into the house to rescue my children – that given the way things are with me motivationally speaking at *t*-1, and given my children's predicament, for example, there is simply no possibility whatever that I will not run in and attempt to rescue them at some time in the vicinity of *t*. But suppose it were also true that given the way things are at *t*-1, there is also no possibility that I shall not run in at precisely 10 mph, along precisely the trajectory I in fact take, through the precise door and at the precise time that I do in fact enter the house, making all the precise individual movements that I do in fact make, at precisely the times that I do in fact make them, etc. Then my suggestion is, we would not be able properly to conceive of what had occurred as a sequence which was truly a sequence of *activity* on my part. The concept of an agent is the concept of something that certain things can be *up to*; the concept of a being which can *settle*, at the time of action, how certain things in the world are to be, in particular, in the first instance, certain things which concern the disposition and movement of parts of her own body. And the concept of an action, I would maintain, is just the concept of a settling of some of these settleable questions by an agent. Not everything about the action has to be up for settling at the time of action – there may be respects in which what will happen is already settled by the time of action – for example, given my motivations and the fact that my children are in a burning house at *t*-1, it may already be settled that some sort of running-into-the-house on my part is going to happen shortly. But *some* things have to remain unsettled, I claim, if what is to occur is to be an exercise of agency on my part at all. Some of the following sorts of question, for example, have to be not yet settled: precisely which movements I shall make; whether I shall go through the window or the door; whether I shall call out as I run from room to room; whether I shall search the kitchen before or after I search the living room; when precisely the action will occur, etc. For if the precise description of the entirety of a course of action is all settled in advance by the past and the laws, there seems no sense in which its agent can count as the true source of any of that action's outcomes, and hence, no possibility that what brings about those outcomes should be an agent's *acting*. The agent would simply dissolve, under such circumstances, into a location, a place where the relevant inexorable events occur. What we can say, then, is this: that any φ-ing which is an agent's action must have some detailed and specific description as a D-ing, say, such that the agent could have refrained from D-ing. For unless this is the case, there is nothing left for the agent to settle at the time of action, and hence no possibility that what occurred at that time should have been an action at all. It is this, I believe, that constitutes the legitimate truth lurking behind the Principle of Alternate Possibilities. Actions being settlings, some of their features and characteristics must be left for the agent to settle at the time of action. And that is as much as to say that actions cannot be determined events. The reason why moral responsibility is inconsistent with determinism, then, is this: the existence of moral responsibility (for anything) requires the existence of agents who may be held responsible; and the existence of agents requires the falsity of determinism.

5. The objection from ignorance

Does this view not imply, though, quite absurdly, that we do not know whether there are any agents? Only if we accept the premise that we do not know whether or not determinism is true. There is a tendency in much recent literature for philosophers to accept that the question whether determinism is true is a scientific question, and that it behoves us to maintain with respect to it that openness of mind that one ought to maintain about questions which it is the job of science to settle. But I should insist rather that the question whether determinism is true is quite plainly a *metaphysical* question, and that philosophical reflection, including, of course, philosophical reflection on the biological phenomenon that is agency, is therefore needed to settle it. In particular, it is not, as many suggest, merely a question for *physicists*, since determinism is a doctrine not merely about the events that come within the purview of physics, but about *all* events, and so one must, at the very least, take a metaphysical view about the dependence of all events on those that belong to physics before the question whether or not determinism is true could even begin to seem as though it might be up for decision within physics. For instance, Fischer claims in a recent work that the doctrine of causal determinism states that:

> for any given time, a complete statement of the (temporally genuine or non-relational) facts about that time, together with a complete statement of the laws of nature, entails every truth as to what happens after that time.
>
> (Fischer 2006: 5)

155

But a complete statement of the facts about any time must presumably include the biological, psychological, sociological and economic facts, as well as the physical ones – and so it is not *immediately* clear why the truth or otherwise of determinism should be a matter only for physicists to decide. Many believe, of course, that facts of all these various sorts supervene on the physical facts. But first, that is not, in itself, a doctrine of physics, but of metaphysics; and second, even if it is true, it is not clear what follows from it, so far as the claim that the question of determinism is a matter for physicists to decide is concerned. It would seem to follow, admittedly, from supervenience that the truth or otherwise of determinism generally rests on the truth or otherwise of *physical* determinism. But it is not clear what our intellectual attitude ought to be to this fact. One view might be that we must await the verdict of physicists before we can hope to decide whether or not the universe is deterministic. But one person's modus ponens is another's modus tollens, and so another view might be this: that since we know from reflecting on what we know of the world in general (including the fact, for example, that it contains agents like ourselves) that determinism quite generally is not true, we know *already* that physical determinism cannot be true. And in case this seems like sheer hubris, just reflect for a moment on how much hubris is implicit in the idea that *physics* alone might be what determines the occurrence and precise nature of such things as wars, political decisions, banking crises, epidemics, the distribution of poverty in a society, the reach of democracy on a continent, the composition of individual works of art and literature. If we were not already mesmerized by a Laplacian picture of reality, the claim that this is so would surely strike us as the height of absurdity.

Of course, it cannot be denied that we cannot rule out for certain that physicists might one day show that physical determinism is true – that each physically characterized state of the world necessitates the next. But it does not follow from this that it is absurd to maintain a view according to which, if physical determinism should turn out to be true, agency would be found not to exist. Instead of this, we might surely say the following: that since we are in fact entitled to a very high degree of assurance about the fact that there are agents, we are entitled to a similarly high degree of assurance that determinism is false. Agency refutes it. It is an absolutely basic part of our worldview that by means of their actions, agents settle a range of matters that were hitherto not settled – and how could this be unless determinism were false? We know, then, with a perfectly reasonable degree of assurance that it *is* false. Of course, here as elsewhere in philosophy, the possibility of radical scepticism remains – perhaps it might be maintained that we cannot know for certain that determinism will not turn out to be true, and so that we cannot know for certain that there are any agents. But the scepticism implied here is radical indeed – as radical as wondering about the existence of the external world, or of other minds, say. Agency is part of our fundamental conceptual scheme. It cannot be ruled out that our fundamental conceptual scheme is mistaken, of course. But how much more likely it should seem to the unprejudiced that we might have made a mistake in formulating the metaphysics that seems to rule it out.

References

Alvarez, Maria (2009) Actions, thought experiments and the 'Principle of Alternate Possibilities', *Australasian Journal of Philosophy*, 87: 61–82.

Dennett, Daniel (1971) Intentional systems, *Journal of Philosophy*, 8: 87–106.

Dennett, Daniel (1987) *The Intentional Stance*, Cambridge, MA: MIT Press.

Fischer, John Martin (1994) *The Metaphysics of Free Will*, Oxford: Blackwell.

Fischer, John Martin (2006) *My Way*, Oxford: Oxford University Press.

Frankfurt, H. (1969) Alternate possibilities and moral responsibility, *Journal of Philosophy*, 66: 829–39.

Frankfurt, H. (1982) 'The importance of what we care about, *Synthese*, 53: 257–72, reprinted in Frankfurt (1988): 80–94.

Frankfurt, Harry (1988) *The Importance of What We Care About: Philosophical Essays*, Cambridge and New York: Cambridge University Press.

Ross, W.D. (1924) Introduction to *Aristotle's Metaphysics* Vol. 1, Oxford: Clarendon.

Sartre, J.-P. (1958) *Being and Nothingness*, trans. Hazel E. Barnes, London: Methuen.

Steward, Helen (2008) Fresh starts, *Proceedings of the Aristotelian Society*, 108: 197–217.

Steward, Helen (2009) Fairness, agency and the flicker of freedom, *Nous*, 43: 64–93.

10
Substance dualism and its rationale

HOWARD ROBINSON

1. What is dualism?

Dualism in the philosophy of mind is the claim that there are two radically different kinds of property (and, hence, state) involved in the constitution of a normal human being, namely physical properties (and states) and mental properties (and states). It may go further, and claim that these different properties are accompanied by different kinds of substance, material substance ('matter') and mental substance (which lacks an uncontroversial label, but which is often identified with consciousness). The crucial point of this claim is that the mental and the physical are, in some way, essentially different; in particular, mental properties and states are not a subclass of physical properties and states. If there is this radical divide between the mental and the physical, we need to know in what this essential difference consists.

The most fundamental difference between the mental and the physical derives from the fact that the core cases of mental states are states of consciousness, and conscious states seem to possess a fundamental property not possessed by physical states. The physical is public, in the sense that, in principle, any physical state is equally accessible (perceivable, knowable) by any normal subject. Any normal person in the right position can, for example, see a brick wall, feel its solidity, or hear the sound of a bird. And though some states and objects are not directly perceivable at all – an electron and its spin, for example – everyone has the same degree of access to it. Anyone can see the trace in the bubble chamber and – at least in principle – anyone can grasp the theory which enables one to interpret the traces that one sees. Conscious states are essentially different, because the subject to whom they belong – and only that subject – has a privileged access to them. I may be able to tell that you are in pain from your behaviour, but only you can feel the pain. It is this latter that gives the most secure knowledge: the behaviour may be feigned, the feeling cannot be. This fundamental difference between the mental and the physical is brought out by the following thought experiment. Imagine someone born deaf. He would not know *what it was like* to hear sounds – C-sharp on a piano, for example. If he came to know all the physical facts about the hearing process – how sound waves work, the operation of the ear and the neurology of hearing in those who can hear – he would still not know *what it is like* to hear, and, consequently, he would not know the experiential or phenomenal nature of sound. The subjective dimension – the dimension of consciousness – seems to escape the third-personal perspective which delivers us the specifically physical facts. These facts, and the states and properties they comprehend are, therefore, not physical properties and states.[1] If our deaf subject were given his hearing through a miracle of modern science, he would come to know something which he had previously not known, however complete his scientific grasp on hearing, namely what C-sharp actually sounds like.

The physicalist also has problems with the nature of thinking. Thought is paradigmatically but not always conscious. When conscious it has the features described above, but, whether or not it is conscious, it possesses a feature which is not possessed by sensations but which is additionally problematic for the materialist monist, namely that thought is always *about* something. This property of 'aboutness' is dubbed *intentionality*. Physical states do not seem to be naturally *about* things: they are just *there* and possess causal powers. Thoughts can even be about things that do not exist (for example, Zeus or Pegasus), but there are no physical relations that

---

[1] I have stated this argument – generally known as the *knowledge argument* – in the way in which it is normally stated, which implies that it relates only to the secondary qualities, treated as subjective states or *qualia*. In fact the same principle applies to our knowledge of primary qualities as to our knowledge of secondary. Our conception of visual shape is as dependent on the subjective dimension of experience as is our conception of red. I develop the consequences of this for materialist strategies in Robinson (forthcoming). The classic modern source for the knowledge argument is Jackson (1982). Jackson and the following discussion take the case, not of a deaf subject, but the rather more contrived case of someone who is prevented from seeing any chromatic colours, but the point is the same. For extended literature on the subject, see Ludlow *et al.* (2004), which also contains Jackson's recantation of his previous commitment to the knowledge argument.

involve non-existents. There have been many attempts to 'naturalize' intentionality, analysing it in causal terms, but none adequately to capture the way that thought captures the world for the subject. (I pursue this point further in Section 5.)

2. Challenges to dualism

Much of the philosophy of mind in the last hundred years – and especially intensively in the last fifty – has been concerned with the attempt to avoid the dualism to which these arguments seem to lead and provide a materialist account of these phenomena. Why has dualism been so strongly resisted? The overall reason is that a monist or unified account of the world has a great appeal simply because it is a unified account, but there are two more specific reasons for its attractiveness. First, the tremendous success of the physical sciences has made it seem that these sciences – especially physics itself – should be able to provide an account of everything in the world: in the contemporary jargon, the world is 'closed under physics' – that is everything that happens is, at bottom, explained by the laws of physics. Now, this can be reconciled with dualism if one is an epiphenomenalist, and denies that the mental ever affects what happens in the physical world, but this is a very counterintuitive and desperate position: it is bizarre to deny that my visual experience, my pains and my thoughts have any influence on what I do.[2] So the success of science is held to tell against dualism. Dualists often respond that this favouring of science is just an ideological prejudice and cannot override the plain phenomena.

Second, there is the problem of interactionism, which is the task of explaining how the mental and the physical are causally related. The simplest objection to interaction between them is that, in so far as mental properties, states or substances are of radically different kinds from each other, they lack that communality necessary for interaction. It is generally agreed that, in its most naive form, this objection to interactionism rests on a 'billiard ball' picture of causation: if all causation is by impact, how can the material and the immaterial impact upon each other? But if causation is either by a more ethereal force or energy or only a matter of Humean

[2] Many neuroscientists developing the work of Benjamin Libet (see Haggard, this volume) claim that there is empirical evidence for epiphenomenalism. This, however, is contentious. (See Bayne, this volume.)

constant conjunction – that is, of the mere regular correlation of events together, one following the other – there would appear to be no problem in principle with the idea of interaction of mind and body. On the Humean model of causation, 'interaction' would require no more than that certain mental events were regularly followed by certain neural events, and certain neural events regularly followed by certain mental events.

Even if there is no objection in principle, there appears to be a conflict between interactionism and some basic principles of physical science. For example, if causal power was flowing in and out of the physical system, energy would not be conserved, and the conservation of energy is a fundamental scientific law. Various responses have been made to this. One suggestion is that it might be possible for mind to influence the *distribution* of energy, without altering its quantity. Another response is to challenge the relevance of the conservation principle in this context. The conservation principle states that 'in a causally isolated system the total amount of energy will remain constant'. Whereas '[t]he interactionist denies . . . that the human body is an isolated system', so the principle is irrelevant (Larmer 1986: 282; this article presents a good brief survey of the options for the interactionist). Some philosophers and some physicists (e.g. Stapp 1993) believe that the indeterminacy of quantum phenomena gives a space for interaction, whereas others entirely deny this. The debate on the plausibility of interactionism has reached no clear consensus, but recently Robin Collins has forcefully pointed out that, according to modern science, energy is not conserved in general relativity, in quantum theory, or in the universe taken as a whole: the idea that conservation is physically ubiquitous is a remnant of nineteenth-century science (Collins 2011).

Nevertheless, if a plausible materialist theory could be provided, its appeal would be very strong. The core and foundation of the twentieth century's attempt to formulate materialism about the mind has been a form of behaviourism or one of its derivatives – that is, consciousness and thought are either identified with tendencies to behave in certain ways, or with internal – probably computational – processes that count as mental because of their function in producing behaviour. The literature is vast, and the theories put forward are often sophisticated and subtle, but the feeling remains amongst many and probably most philosophers – even ones who aspire to be materialists – that these approaches cannot capture the subjectivity of consciousness, or the strange capacity of thought which enables us to assimilate the world into our understanding. There is a widespread feeling that the reasons I gave initially for being a dualist are more

161

sidestepped than properly addressed by behaviouristic or functionalist theories.[3]

3. Why be a substance dualist?

I have argued that, despite the appeal of materialism, there are good reasons for being a dualist. But one might think that one should stay as near to materialism as possible, and be a property dualist. The brain is, after all, a very complex object, and perhaps for good evolutionary reasons it has developed these strange and unique properties. One might conclude, therefore, that these properties are properties *of the brain* and, hence, *of the human animal*, not the properties of some strange immaterial substance. The conscious mind, on this account, is just the integrated bundle of the human animal's conscious mental states – 'integrated' by their dependence on the sophisticated organization of the brain.

There are many philosophers, however, running from Plato, through Descartes, and including contemporaries such as Richard Swinburne (1986) and John Foster (1991), who do not accept that the mind is just a bundle of properties associated with a human body: they think, instead, that it is an immaterial *substance* in its own right. These are the *substance dualists*.[4]

The first question that arises is 'What is substance dualism? What is involved in categorizing the mind as a substance in its own right?' This, of course, raises the whole issue of what a substance is, and this is far too large an issue to deal with en passant.[5] For present purposes, we can stay with the intuitive contrast between objects and their properties. Objects are relatively self-standing entities – they do not exist *by* belonging to something else, whereas properties only exist *by* being the property of some-

---

[3] The classic statements of these theories are, for behaviourism, Ryle (1949) and for the causal theory, Armstrong (1968). Putnam (1975 [1960]) brought in the computational analogy and the term 'functionalism', which came to dominate the discussion, but it is Armstrong who most comprehensively and lucidly captures the behavioural/causal approach to the mind. For a thorough exposition and discussion of the inadequacies of this approach, see Chalmers (1996) and Robinson (2009 [1982]).

[4] This is not strictly accurate. Not everyone who believes that the mind is an immaterial substance is a dualist. Some, like Foster (and the present author) are idealists. But, for present purposes, it is the status of mental substance that is relevant.

[5] For anyone interested in the general issue, see the entry 'Substance' in the *Stanford Encyclopaedia of Philosophy* – Robinson (2004/9).

thing, usually an object. Objects are, of course, usually causally dependent on other objects in various ways, but they are not conceptually dependent on them, in the way that, for example, the roundness property of the ball depends on the existence of the ball. Objects, in this sense, are pretty well what is meant by 'substance'. So the mind is a substance if it is not just a collection of properties of the body or brain, and if its dependence on the body or brain is mere causal dependence; it is not conceptually dependent. There are many issues that could be raised here, but this must suffice for present purposes.[6]

Descartes' most famous reason for thinking that the mind was a substance in its own right occurs in the *Second Meditation* (Descartes 1984 [1641]). It is that we can imagine that our current conscious experience be just as it is, yet, because we are being deceived by an 'evil demon', none of the physical things we seem to experience, including our own bodies, are genuine. We could, at this moment, be in a completely hallucinatory or virtual reality, with all the physical things we seems to experience being unreal, including our own bodies. Descartes thinks that the fact that we can conceive this to be the case shows that the mind is a distinct thing from the body and can, in principle, exist without the body. If you can imagine one thing existing without another, then the latter cannot be essential to the former and they are distinct entities. So the mind is a distinct entity from the body.

Although this argument has a certain appeal and there are still philosophers who think it has force, it is also deeply immersed in many controversies. The first of these concerns the move from something's being imaginable to its being really possible. Modality – the theory of necessity and possibility – is a very difficult and murky area, but there are philosophers who hold, that, though imaginability is not a complete guarantee of possibility, it is a very good – possibly the best available – guide to it (e.g. Hart 1988). There are many others, however, who think that, at least in controversial matters, it is a very weak indicator. Another problem concerns the conception of mind that seems to lie behind Descartes' argument. This conception of mind is often referred to as the 'Cartesian theatre' conception, and according to it the life of the mind is wholly internal, a

---

[6] The argument is, of course, much more complicated than I suggest. You could argue that objects are conceptually dependent on properties both because there cannot be an object without properties and at least some of these properties may be essential to the object. For a more thorough discussion, see Robinson (2004/9).

kind of bubble of consciousness confined beyond and 'further inside' than the brain. Only if you take this view of conscious life, it is argued, can you imagine consciousness to be as it is and the physical world taken away. This 'Cartesian theatre' view of the mind is disputed by direct realists, who believe that our conscious life directly encompasses the external world, especially our own bodies. If this latter is the real situation, they argue, then you cannot abstract the physical world and leave experience untouched, for external physical objects constitute – and do not merely cause – the contents of our consciousness.[7]

Because of the depth of controversy surrounding Descartes' argument, that is not the argument for substance dualism on which I wish to concentrate. The line I prefer to investigate is connected with theories of personal identity and rests on the belief that bundle dualism (the theory that the mind is not a substance but only a collection of immaterial properties or states) cannot accommodate certain essential features of personal identity – what makes a person the particular person that he or she is. I made some elementary remarks about the notion of *substance* above. The argument I am about to consider is ambitious because it is intended to prove not merely that the mind is a substance, but that it is a *simple* substance – one of the world's atomic entities, though in a rather special sense.

One must approach this topic by first considering the identity conditions for ordinary physical objects. Most usually, this is considered by asking what makes an object the same object over time. So let us start with a particular object – say a rather old-fashioned wooden ship. This is a classic example – the ship of Theseus. Theseus, being a responsible sailor, has his ship repaired over time, so that finally every plank in its original construction has been replaced by a new one. Being a sentimentalist, however, he preserved all the old planks, and when every one had been replaced, he reassembled the old ones into the form of the original ship. The classic question is – Which is the same ship as the original, the one that has evolved or the one that is constructed from the original matter? There are philosophers who think that there must be a specific and true answer to this question, but probably most would agree that it is a matter of what you are interested in – whether you are an antiquarian or commercial sailor – there is no 'true' answer, it is a matter of convention or decision. Once you have

---

[7] This is one version of the fashionable doctrine called *externalism about mental content*. See, e.g., McDowell (1994).

told the story as I have told it, those are all the facts: which is 'really' the same ship is not a genuine further question. There are many similar cases. Once you have told the narrative of what happened to England and its governance between 1000 and 1100, the question of whether it was 'really' a different country after 1066 is not a further question beyond all the changes and continuities that can be recounted in their own right. Speaking loosely, one can say that the identity of physical objects over time is a matter of degree. ('Loosely' because it is generally held that, strictly speaking, things cannot be 'partially identical' any more than something can be 'fairly unique'. But in both cases one can see what is being said.)

There is a long tradition, dating at least from the eighteenth-century Scottish philosopher Thomas Reid (1969 [1785]), for arguing that the identity of persons over time is not a matter of convention or degree in the way that is the case for the examples we have just been considering. There is something absolute – all or nothing – about one's being numerically the same person at 70 as at 7.[8] Unfortunately this intuition is controversial and does not command universal, or even general, assent. Growth, ageing, and especially radical changes in personality due to accidents or diseases are claimed to make one into 'a different person'. Some would say that qualitative differences in personality must be distinguished from numerical identity as the same person, others, that there is no absolute difference between these two, as there is not in the identity of England through invasion and radical political change.

I think that the issue can be made sharper and clearer, however, if one moves from considering identity through time to the rather less familiar matter of identity under counterfactual circumstances, especially those concerning origin. Instead of asking whether Theseus's ship was the same object when half its planks had been replaced, we ask whether it would have been the same ship if it had been constructed with different materials in the first place. So we are not considering changes within its life as a boat, but possible differences at its origin. Thus we are considering *counterfactuals of origin*, that is, things that might have been different at the beginning of the existence of an object. (Such things are *counterfactuals* because they state how things *might* have been, not how they, in fact, were.) We would probably agree that if the ship had been made not of wood but of gold, it would not have been the same ship at all. But if it had been made of, say,

---

[8] This view is also defended in Chisholm (1976).

10% different planks and 90% the same . . . ? This thought experiment can be duplicated for any complex physical object and we are once again tempted to reply that there is no fact of the matter about whether it would or would not have been the same in the borderline cases. Once the story has been told about such and such differences, those are all the real facts. There is more or less overlap of constitution, but what, if anything, one says about identity is a matter of choice. As I hope to show, a similar treatment cannot be meted out in the case of persons, when it comes to these counterfactual cases, even though it looked as if it could in the case of identity through time.

Let us try to apply the same thought experiment to a human being. Suppose that a given human individual – call him Jones – had had origins different from those which he in fact had, such that whether that difference affected who he was, was not intuitively obvious. We can approach this by imagining cases where it seems indefinite whether what was produced was the same body as Jones in fact possesses. What would count as such a case might be a matter of controversy, but there must be one. Perhaps it is unclear whether Jones's mother would have given birth to the same human body if the same egg from which Jones's body came, had been fertilized by a different though genetically identical sperm from the same father. Some philosophers might regard it as obvious that sameness of sperm is essential to the identity of a human body. In that case, imagine that the sperm that fertilized the egg had differed in a few molecules from the way it actually was; would that be the same sperm? If one pursues the matter far enough there will be indeterminacy which will infect that of the resulting body. There must therefore be some difference such that neither natural language nor intuition tells us whether the difference alters the identity of the human body; a point, that is, where the question of whether we have the same body is not a matter of fact.

These are cases of substantial overlap of constitution in which that fact is the only bedrock fact in the case: there is no further fact about whether they are 'really' the same object.

My claim is that no similar overlap of constitution can be applied to the counterfactual identity of minds. In Geoffrey Madell's words:

> But while my present body can thus have its partial counterpart in some possible world, my present consciousness cannot. Any present state of consciousness that I can imagine either is or is not mine. There is no question of degree here.
>
> (Madell 1981)

Why is this so? Imagine the case where we are not sure whether it would have been Jones's body – and, hence, Jones – that would have been created by the slightly modified sperm and the same egg. Can we say, as we would for an object with no consciousness, that the story 'something the same, something different' is the whole story: that overlap of constitution is all there is to it? For the Jones body as such, this approach would do as well as for any other physical object. But suppose Jones, in reflective mood, asks himself 'if that had happened, would I have existed?' There are at least three answers he might give to himself: (i) 'I either would or would not, but I cannot tell.' (ii) 'In some ways, or to some degree, I would have, and in some ways, or to some degree, I would not. The creature who would have existed would have had a kind of overlap of psychic constitution and personal identity with me, rather in the way there would be overlap in the case of any other physical object.' (iii) 'There is no fact of the matter whether I would or would not have existed: it is just a misposed question. There is not even a factual answer in terms of overlap of constitution.' I shall discuss (ii) in the rest of this section, and move to answer (iii) at the end of Section 4.

The second answer parallels the response we would give in the case of bodies. But as an account of the subjective situation, it makes no sense. Call the creature that would have emerged from the slightly modified sperm, 'Jones*'. Is the overlap suggestion that, just as, say, 85% of Jones*'s body would have been identical with Jones's original body, about 85% of his psychic life would have been Jones's? That it would have been like Jones's – indeed that Jones* might have had a psychic life 100% like Jones's – makes perfect sense, but that he might have been to that degree, the same psyche – that Jones '85% existed' – makes no sense. Take the case in which Jones and Jones* have exactly similar lives throughout: which 85% of the 100% similar mental events do they share? Nor does it make sense to suggest that Jones might have participated in the whole of Jones*'s psychic life, but in a rather ghostly 'only 85% there' manner. Clearly, the notion of overlap of numerically identical psychic parts cannot be applied in the way that overlap of actual bodily part constitution quite unproblematically can.

There are two things to notice about this argument. The first is how the identity across counterfactuals of origin case differs from that of identity through changes across time. The second concerns the peculiarly strong sense of individuality that goes along with self-consciousness.

The first of these points concerns what one might call *empathetic distance*, which is essential to the problematic nature of identity through time but irrelevant in the counterfactual case.

Suppose that my parents had emigrated to China whilst my mother was pregnant with me, and that, shortly after my birth, both my parents had died. I was then taken in by Chinese foster parents, lived through the revolution and ended up being brought up in whatever way an alien would have been brought up in Mao's China. None of this person's post-uterine experiences would have been like mine. It seems, on the one hand, that this person would obviously have been me, and, on the other, that it is utterly unclear what kind of empathetic connection I can feel to this other 'me'. If I ask, like Jones, 'would this have been me?', I am divided between the conviction that, as the story is told, it obviously would, and a complete inability to feel myself into the position I would then have occupied. This kind of failure of empathy plays an important role in many stories that are meant to throw doubt on the absoluteness of personal identity. It is important to the attempt to throw doubt on whether I am the same person as I would become in fifty years' time, or whether brain damage would render me 'a different person' in more than a metaphorical sense. It is also obviously something that can be a matter of degree: some differences are more empathetically imaginable than others. In all these cases our intuitions are indecisive about the effect on identity. It is an important fact that problems of empathy play no role in the counterfactual argument. The person who would have existed if the sperm had been slightly different, could have had as exactly similar a psychic life to mine in as exactly similar environment as you care to imagine. This shows the difference between the cases I have discussed and the problematic cases that involve identity through time. In those cases the idea of 'similar but not quite the same' gets empirical purchase. My future self feels, in his memory, much, but not all, of what I now feel. In these cases, overlap of conscious constitution is clearly intelligible. But in the counterfactual cases, imaginative or empathetic distance plays no essential role, and the accompanying relativity of identification gets no grip.

4. Individuality and consciousness

The second point to notice is the light it throws on the concept of what the medievals called *haecceitas*. *Haecceitas* translates as 'thisness' and is, according to certain philosophers, the feature of an object which, additional to its ordinary properties, makes an individual thing the particular that it is. Most – though by no means all – philosophers regard this as a very suspect notion: in the case of complex physical bodies, for example, it is difficult to imagine what a *haecceitas* would consist in or how it relates to the other features of the object, and so the suggestion that there is such a thing seems to be pure mystery-mongering. By contrast, in the case of minds we do have a form of *haecceitas* which, in a sense, we all understand, namely our identity as subjects. It is because we intuitively understand this that we feel we can give a clear sense to the suggestion that it would, or would not, have been ourselves to which something had happened, if it had happened: and that we feel we can understand very radical counterfactuals – for example, that I might have been an ancient Greek or even a non-human – whereas such radical counterfactuals when applied to mere bodies – that this wooden table, for example, might have been the other table in the corner or even a pyramid – make no intuitive sense. It is possible to argue that the suggestion that my mind might have been in another body ultimately makes no sense, but it makes a prima facie sense – it seems to have content – in a way that a similar suggestion for mere bodies does not. The very fact that the counterfactuals for subjects seem to make sense exhibits something not present in the other cases, which is available to function in the role of *haecceitas*. Only with consciousness understood in a Cartesian fashion can *haecceitas* be given an empirical interpretation.

The reflections on *haecceitas* can be developed further. If one thinks of a true particular as being something that can sustain counterfactuals and still be clearly the same individual, perhaps minds are the most genuine particulars that there are. We have seen that there is vagueness in the cases of identity for complex physical objects. This, we suggested, makes such identity a matter of convention or decision and not a true matter of fact. Reverting to our original example, the statement 'there is a ship of a certain sort and composition' gives you all the real or fundamental facts out in the world; the situation is adequately characterized by a statement which is a generalized or quantified one –'there is *some* such and such' – one can dispense with more exact identificatory discourse in this context.

There might, however, be the following response. It might be argued that what I say is true of composite objects but not of the units that compose them. In the case of Theseus's ship, these are the planks, and obviously the same problem about their counterfactual identity can be raised as for the ship itself, for the planks are composites, too. We can ask 'would it have been the same plank if it had been composed of such and such different atoms?' and the same problem arises. One might think that this shows that composite physical objects have a precarious hold on identity and, therefore, on existence as real individuals, but that for true atoms the same problem cannot arise. But atoms, too, are not problem-free under counterfactuals of origin. Suppose a particular electron (pretending electrons to be basic for purposes of illustration) came into existence at a particular point in space time. If it had instead come to be in a slightly different location, would it have been the same one? Always questions of this sort can be generated for physical objects to show that there is no real or ultimate difference between qualitative similarity and real identity as particulars in their case. Only the inwardness of subjectivity can deliver the difference. It is not simply the simplicity and unity of the self that constitutes its existence as a true individual, but its nature as conscious, from which that unity derives.

Some philosophers of physics have asserted that *haecceitas* must be attributed to certain sub-atomic particles – bosons, for example – if they are to be regarded as proper individuals.[9] But it is agreed that quantum theory itself does not require quantal entities to be individuals – indeed, it makes as much sense, if not more, if they are treated as something less than individuals. It is not possible here to enter into a debate about how best to interpret quantum theory, though it seems to me that those who wish to endow bosons with *haecceitas* do so only because they believe that this is necessary to endow them with the degree of individuality possessed by more ordinary physical objects. As I am sceptical about the status of all physical objects as full individuals, this motive gains no hold. Nevertheless, none of the arguments that I have presented for treating the self as a simple individual are in any way undermined if physical simples are also individuals.

Someone who, unlike the boson *haecceitist*, was impressed by my argument that atomicity itself is not enough alone to guarantee status as a genuine individual might be led to look again at the case of the self. We considered the option that its identity might be a matter of degree, and rejected this. But what about the suggestion that there is no firm difference between qualitative similarity and numerical identity? We have a strong feeling that there must be a difference between these two in the case of bodies, yet this seems to be mistaken. Could our sense that there must be such a distinction in our own case also be an illusion? Is that conception of the self which makes us feel so sure that someone psychically just like me but with a somewhat different origin either is me or is not something that

[9] For a lucid account of these issues, see French (2006).

needs 'deconstructing', after the fashion of Derrida or Nietzsche or Hume? That this is so is the substance of (iii) above.

I do not think that the idea 'just like me but the idea of whether it would be me or not has no content' can be made acceptable. Whereas in the case of physical objects we can see, after a little thought, that though the qualitatively similar gives us all we thought we needed by talking about particulars, it will not do this in the case of minds.

Consider the following example. Suppose you discover that, in the very early stages in the womb, you were one of twins, but that the other did not develop, and that it could have easily happened the other way round; the other would have survived and you died in the first few days. The similarity between you as survivor and your twin, had he survived in your stead, both in genetic endowment and environmental circumstances and subsequent experience, could have been almost complete. Nevertheless, there is no sense that, on reflection, it makes no serious factual difference, concerning your own fate, which of the two survived. Just as it is true that, if your parents had never met, then you would not have existed, equally, if the other bundle of cells had developed instead of yours, you would not have existed. This is, in no sense, a matter of decision, convention or degree.

It would seem that the only possible answer to the question which I supposed Jones to have asked himself on page 167, 'if that had happened, would I have existed?' is (i), 'I either would or would not, but I cannot tell'. If there is a real fact, independent of our convention or decision, in this case, then it shows that counterfactual identity facts are real facts in the case of minds, in a way that they are not for physical objects.[10]

5. Making sense of the substantial self . . .?

The argument just presented has, I think, great intuitive appeal, but, on the other hand, it also seems to leave many puzzles. The first concerns the

---

[10] I have not discussed or allowed for David Lewis's notorious modal realism. According to Lewis, every possibility represents a completely different spatio-temporal system. So the sentence 'I might have had a fried egg for breakfast yesterday' (though I did not) is made true by the existence of a universe spatio-temporally unrelated to this one in which someone otherwise just like me (my 'counterpart') did have a fried egg for breakfast on the parallel day. On this view, in the most basic sense, *nothing at all* sustains counterfactuals, because all other possibilities are realized in counterpart entities, which, strictly speaking, are different things from the objects in the world we inhabit.

nature of mental or immaterial substance itself: what is its 'essence'; how should one characterize it? More often than not, the answer is given that it is consciousness. Some feel unhappy with the idea that consciousness can be treated as a special kind of 'stuff'. Even if one puts aside that issue, what are we to say about the existence of the self or mind during periods of unconsciousness, such as deep sleep or anaesthesia, if its nature is consciousness? One way of avoiding this difficulty is that taken, for example, by Galen Strawson (2009). Strawson thinks that a particular self lasts roughly only as long as a single span of attention, so a given person will be composed of many selves over time. If one is to avoid this and similar counterintuitive ways of dealing with lapses of consciousness, how can one do it? I want to give a hint on how one might approach this problem by discussing a second apparent difficulty for the theory that the self is a simple substance.

The second problem is as follows. The argument of the previous sections attributes the conscious subject a unity and simplicity unique in nature. It is unique because I have cast doubt on the existence of any true individuals that are purely physical. But putting aside the claim that there are no true atomic physical individuals, we still have a problem about the atomic and simple nature of the self: how can something as complex as a human subject be a simple entity? People have a variety of faculties and capacities, and an almost unlimited number of memories, beliefs, desires, etc. What does it mean to say that such an entity lacks parts or composition? Attempts to answer this question are liable to drive one into what Russell somewhere described as 'soupy metaphysics' and I cannot venture too far into such territory here.[11] Some insight into how one might approach the problem can be gained by considering the 'unity in diversity' that is an essential feature of thought.

Peter Geach has argued that the 'activity of thinking cannot be assigned a position in the physical time-series' (1969: 34). His reason for this is that, though the expression of a thought using a sentence will be spread through ordinary time, one's grasp on the content must come as a whole. If it did not, then by the time one had reached '1066' in the sentence 'the battle of Hastings took place in 1066' one's consciousness of the other components of the thought would have passed into history. What the sentence expresses as a whole is the thought of which one is conscious. Something that has an

---

[11] For a 'non-soupy' defence of simplicity rather different from mine, see Chisholm (1991).

essential unity finds expression in something that is complex. The position seems thus to be the following. The expression of a thought in a sentence is spread out in the normal 'flowing' empirical time. But the thinking of the thought which, in some sense, 'lies behind' (but not necessarily temporally before) this, is not temporally structured in the same way. Something which is implicit in the thought is laid out explicitly in the sentence. One experiences a thought *in* a sentence – or sometimes in other, non-verbal, images – but as a unity that a mere string of sounds or images does not possess.

Isn't this a somewhat mysterious doctrine? It is, but it is true to the phenomenology of thought, and if one tries to avoid a conclusion along these lines one reaches even more counterintuitive conclusions. Consider, for example, the attempt to demystify thinking by treating it as a computational process. Jerry Fodor (1975, 1979) treats consciousness as irrelevant to thought, which is a computational process carried out in the purely formal 'Language of Thought' (LOT) in the brain. This leads to certain serious limitations. For example, any term not definable within the system must be primitive and innate to it. Because very few terms in natural language are explicitly and precisely definable, this leads Fodor to claim that the LOT equivalents of almost all terms, such as, for example, 'xylophone' and 'crocodile', must be innate. Furthermore, formal systems cannot 'upgrade' themselves and so there is no natural development in the power of the system. Any purely physical 'thinking' machine must be what Daniel Dennett (e.g. 1981) calls a *syntactic engine,* which means that, like a computer, it works solely from the physical properties of the symbols it manipulates and not through any grasp of meaning. This appears to leave out anything we do in the way of responding to the meaning of what people say or unpacking meanings or being genuinely inventive: a purely syntactic engine – one in which *understanding* has no essential role – it would seem cannot do these things.[12] As Mark Baker (2011) points out, this lacuna in the scientific explanation of thought had already been indicated by Chomsky, fifty years ago. Chomsky divided language into three elements, the lexicon, syntax and the Creative Aspect of Language Use:

---

[12] There is, of course, a massive literature on these topics. A defence of a perspective similar to mine but from a scientific point of view can be found in Penrose (1994). Several physicalist philosophers, most notably Dennett, respond to the features of thought which are difficult for them to accommodate by treating the semantic features of thought as the creation of interpretation from a third person perspective. For the problems with this, see Robinson (2010).

science, he claimed, had nothing to say about the last. This is, in part, at least, because understanding is not driven by syntax alone. This would suggest that the Creative Aspect of Language, and, hence, the development of thought, when that involves more than formal inferences, but depends on our grasp of meanings and our understanding of our own projects, depends on something more than features of the neural/computational machinery. The natural candidate for this source of creativity is the self. Furthermore, for long-term tasks, such as the development of our life projects, Strawson's ephemeral selves would not be adequate. When I follow out an argument, let alone develop my view of the world, I am not trying to convince my heirs, but myself. It would require very coercive reasons to make me abandon the idea that this is the project of a single self and subject and Strawson does not have such arguments.[13]

This does not tell us directly, however, how an essentially conscious entity can survive periods of unconsciousness. Geach's remarks about thought, however, do give us a lead by casting doubt on our common-sense view of time and temporal relations. Geach also says:

> The difficulty felt over saying that a thought need be neither long, nor short, nor instantaneous comes about, I suggest, from a (perhaps unacknowledged) assumption of a Newtonian or Kantian view of time: time is taken to be logically prior to events, events, on the other hand, must occupy divisible stretches or else indivisible instants of time. If we reject this view and think instead in terms of time-relations, then what I am suggesting is that thoughts have not got all the kinds of time-relations that physical events, and I think also sensory processes, have.
>
> (Geach 1969: 36)

The expression of thought is a process taking time, the thought itself is not. If you think – like, in their different ways, both Newton and Kant did – that time is an all-encompassing necessary framework of events, then the thought and its thinker must be located somewhere specific within the ubiquitous time series that also houses the expression of the thought. But if what we think of as physical time is a construct from the relations between certain kinds of events and objects, and if certain other objects and occurrences are related in very different ways from the time-constructing events, then these different ones will not be in what we think of as physical time. In so far as the apparent complexity of the self flows from the

---

[13] There are other cogent objections to ephemeral selves. Richard Swinburne (1997), for example, argues that this is a less simple theory than that of the enduring, single self.

diversity of its actions within time, we can see how it may be a genuine unity yet express itself as a diversity. We can also see how we might approach the first problem we raised for the substance dualist, because these remarks about time give us a way of understanding how a single consciousness might have what is apparently an intermittent existence with empirical time. That time is (like space) a construct from the experiences of subjects and the subject is not itself within it.[14]

6. Substance dualism and free will

The topic of this volume is freedom of the will, and the role of my exposition and defence of substance dualism is more to provide a useful background than to contribute directly to the topic. There is, however, some direct relevance, which I shall briefly outline.

Free will is a property (if of anything) of persons and the issue of what a person is is, therefore, relevant. If physicalism is true and a person is just a living human body, then the libertarian account of freedom would seem to be ruled out: events in physical systems can be rigidly caused or to some degree random, but a physical event can hardly be free in the libertarian sense. So libertarian freedom requires dualism of some kind. But a dualism simply of properties and events – a form of bundle dualism – would not seem to be adequate for libertarianism, for libertarianism requires a substantive conception of agency, which a mere bundle of co-conscious events would not seem to provide. Given that the human animal is not the right kind of thing, on its own, to be a person, or a libertarian free agent, and mere mental events do not seem fitted for agency in a strong form, a more substantial form of dualism seems to be necessary. Furthermore, ephemeral selves of the kind postulated by Strawson (who rejects libertarian freedom) would not support a classical sense of responsibility, for it should be the same agent who acts and who is held responsible, later, for his actions. So the substantial self should exist throughout the whole period for which someone can be held responsible. It is clear, I think, then, that a classic libertarian freedom, and a classic conception of agency and responsibility depend on substance dualism. Other accounts of freedom, of course, may make less specific demands on our theory of the self.

---

[14] I discuss this at length in Robinson (2007).

7. Conclusion and warning

I must repeat what has been said or hinted at in various places in this essay, namely that almost everything that I have said that goes beyond mere description of positions is highly controversial. Everything that I discuss has been a subject of massive controversy in recent years and much of it has been contentious since philosophy began.[15] Substance dualism has been an unfashionable – even despised – position ever since Ryle's crude polemic (1949) in *The Concept of Mind*. Few of my professional colleagues would be inclined to endorse the positive arguments that I have offered, and, when it comes to the speculations in Section 5, there would be more raised eyebrows than applause. Nevertheless, I think that, if we are to make sense of the idea that the self or the mind is an immaterial substance, a path somewhat like the one I have set out is the way we must follow.

References

Armstrong, D.M. (1968) *A Materialist Theory of the Mind*, London: Routledge & Kegan Paul.

Baker, M. (2011) Brains and souls: grammar and speaking. In M. Baker and S. Goetz (eds) *The Soul Hypothesis*, London and New York: Continuum.

Baker, M. and Goetz, S. (eds) (2011) *The Soul Hypothesis*, London and New York: Continuum.

Chalmers, D. (1996) *The Conscious Mind*, Oxford: Oxford University Press.

Chisholm, R. (1976) *Person and Object: A Metaphysical Study*, La Salle, IL: Open Court.

Chisholm, R. (1991) On the simplicity of the soul, *Philosophical Perspectives*, 5: 157–81.

Collins, R. (2011) The energy of the soul. In M. Baker and S. Goetz (eds) *The Soul Hypothesis*, London and New York: Continuum.

Dennett, D. (1981) *Brainstorms*, Cambridge, MA: MIT Press.

Descartes, R. (1984 [1641]) Meditations on first philosophy. In J. Cottingham, R. Stoothoff and D. Murdoch (eds) *The Philosophical Writings of Descartes, Vol. II*, Cambridge: Cambridge University Press, pp. 1–62.

Fodor, J. (1975) *The Language of Thought*, Hassocks: Harvester Press.

Fodor, J. (1979) *Representations*, Cambridge MA: Harvard University Press.

Foster, J. (1991) *The Immaterial Self*, London: Routledge.

---

[15] Olaf Stapledon's science fiction novel *Starmaker* follows the history of intelligent life in the universe through the history of time until the universe finally disintegrates. By the end, all interesting problems have been solved by some intelligent life form at some time – except the mind-body problem. Stapledon was a Balliol man, but even that conceit knew its limits.

French, S. (2006) Identity and individuality in quantum theory. In *The Stanford Encyclopaedia of Philosophy*, http://plato.stanford.edu/entries/qt-idind/

Geach, P. (1969) What do we think with? In *God and the Soul*, Cambridge: Cambridge University Press, pp. 30–41.

Hart, W.D. (1988) *Engines of the Soul*, Cambridge: Cambridge University Press.

Jackson, F. (1982) Epiphenomenal qualia, *Philosophical Quarterly*, 32: 127–36.

Larmer, R. (1986) Mind-body interactionism and the conservation of energy, *International Philosophical Quarterly*, 26: 277–85.

Lewis, D. (1986) *On the Plurality of Worlds*, Oxford: Blackwell.

Ludlow, P., Nagasawa, Y. and Stoljar, D. (eds) (2004) *There's Something about Mary*, Cambridge, MA: MIT Press.

Madell, G. (1981) *The Identity of the Self*, Edinburgh: Edinburgh University Press.

McDowell, J. (1994) *Mind and World*, Cambridge, MA: Harvard University Press.

Penrose, R. (1994) *Shadows of the Mind*, Oxford: Oxford University Press.

Plato, *Phaedo*, many editions.

Putnam, H. (1975 [1960]) Minds and machines. In *Philosophical Papers*, Vol. II, *Mind, Language and Reality*, Cambridge: Cambridge University Press, pp. 362–85.

Reid, T. (1969 [1785]) *Essays on the Intellectual Powers of Man*, Cambridge, MA: MIT Press.

Robinson, H. (2009 [1982]) *Matter and Sense*, Cambridge: Cambridge University Press.

Robinson, H. (2004/9) Substance. In *The Stanford Encyclopaedia of Philosophy*, http://plato.stanford.edu/entries/substance/

Robinson, H. (2007) The self and time. In P. van Inwagen and Dean Zimmermann (eds) *Persons Human and Divine*, Oxford: Clarendon, pp. 55–83.

Robinson, H. (2010) Quality, thought and consciousness. In P. Basile, J. Kiverstein and P. Phemister (eds) *The Metaphysics of Consciousness: RIP Supplement 67*, Cambridge: Cambridge University Press, pp. 203–16.

Robinson, H. (forthcoming) Qualia, qualities and our conception of the physical world. In Benedikt Gocke (ed.) *The Case for Dualism*, Indiana: Notre Dame University Press.

Ryle, G. (1949) *The Concept of Mind*, London: Hutchinson.

Stapp, H.P. (1993) *Mind, Matter, and Quantum Mechanics*, Berlin: Springer-Verlag.

Strawson, G. (2009) *Selves: An Essay in Revisionary Metaphysics*, Oxford: Oxford University Press.

Swinburne, R. (1986) *The Evolution of the Soul*, Oxford: Clarendon. Revised edition 1997.

11
What kind of responsibility must criminal law presuppose?

R.A. DUFF

1. The challenge to criminal law

The criminal law presumes that adult defendants who appear in its courts are responsible agents who can be held to account for what they have done, and be convicted and punished if they are proved to have committed the crimes with which they are charged. Given the presumption of innocence, it is of course the prosecution's job to prove the defendant's criminal responsibility: to prove that he committed the offence charged with such 'fault element' – intention, knowledge, recklessness, or negligence – as the law requires; for instance, that he intentionally wounded another person, or that he recklessly damaged another's property, or that he drove without due care and attention. In order to discharge that burden of proof, however, the prosecution does not also need to prove that the defendant is, or was at the time of the alleged offence, a responsible agent who can be properly held to account in this way: rather, his standing as a responsible agent is presumed. It is open to the defendant (or his counsel) to offer a defence of non-responsibility: to adduce evidence that although (or even if) he committed the crime charged, he was at the time suffering from responsibility-negating insanity or automatism; but the burden of introducing and establishing such a defence lies on the defendant.

> every man is to be presumed to be sane, and to possess a sufficient degree of reason to be responsible for his crimes, until the contrary be proved . . . to establish a defence on the ground of insanity, it must be clearly proved that, at the time of the committing of the act, the party accused was labouring under such a defect of reason, from disease of the mind, as not to know the

nature and quality of the act he was doing; or, if he did know it, that he did not know he was doing what was wrong.[1]

Most defendants against whom the prosecution's case is proved cannot plausibly claim to have been suffering from such a 'defect of reason'; most are well aware of the legally relevant 'nature and quality' of their actions – for instance that they are wounding another person, or taking or damaging their property; most are well aware that what they are doing is defined by the criminal law, and by the morality of those around them if not by their own morality, as 'wrong'. The law treats them accordingly as responsible agents, and convicts and punishes them for their crimes.

Suppose, however, that we came to accept the conclusion for which Galen Strawson and others have persuasively argued: that 'we can never be truly or ultimately morally responsible for our actions'?[2] Would (or could, or should) this lead us to doubt the law's presumption of responsibility? If it would (or should) do so, what kinds of change in the criminal law should we then seek, to prevent it relying on such a dubious or untenable presumption?

Strawson himself implies that such doubts should arise only if we take something other than a strictly, and simply, consequentialist view of the aims and values of the criminal law. For such 'true' or 'ultimate' responsibility is, he argues,

> responsibility of such a kind that, if we have it, then it *makes sense* to suppose that it could be just to punish some of us with (eternal) torment in hell and reward others with (eternal) bliss in heaven . . . [Such] responsibility exists if punishment and reward can be fair without having any sort of pragmatic justification whatever.
>
> (Strawson, this volume: 128)

If our concerns in allocating punishments and rewards are purely pragmatic, to do with securing certain kinds of beneficial consequence or

---

[1] The core of the (in)famous M'Naghten Rules laid down for the insanity defence in England by the House of Lords in *M'Naghten* (1843) 10 Cl & F 200: the passage quoted here is from Lord Chief Justice Tyndal's authoritative response (at p. 210) to questions put to the judges by the House of Lords. The content of the Rules, and the burden of proof they lay on the defendant, have been controversial (for a good discussion of the current law, see Simester *et al.* 2010: 701–14): but even in jurisdictions where the defence is defined more generously, the onus typically lies on the defendant at least to adduce evidence of insanity that would suffice, if not rebutted, to create a reasonable doubt about whether he can be properly held responsible; his responsibility, that is, is still defeasibly presumed.

[2] Strawson (this volume: 126); see also Strawson (1994). For a useful relevant response, see Fischer (2006).

averting certain kinds of harmful consequence, we do not need to worry directly about such ultimate responsibility: we must be concerned with 'what works', and what worked before we abandoned our belief in ultimate responsibility can still be expected, for the most part, to continue to work after we abandon that belief. Thus if we find the justification of criminal punishment in its efficiency as a deterrent, we will believe that a penal system can be justified only if potential offenders can be deterred from crime by the prospect of punishment: but the causal efficacy of the threat of punishment clearly does not depend on our having ultimate responsibility; some might indeed argue that it depends on the absence of such responsibility, since it depends on our susceptibility to the causal influence of deterrent threats. Similarly, the absence of ultimate responsibility does not undermine punishment's capacity to incapacitate potential offenders, or to reform offenders (in so far as that was ever a realistic aim). If we take such a purely consequentialist view of punishment, as being justified by its efficiency in preventing crime by deterring, incapacitating, or reforming offenders, then abandonment of any belief that we might have had in ultimate responsibility need not affect our views either of its justifiability or of how it can best be used to serve its crime-preventive ends.[3]

However, few theorists of punishment, and I suspect few people who think about the topic, take such a purely consequentialist view. Some argue that the primary purpose of punishment is retribution for past wrongdoing – to impose on offenders the penal burdens that they deserve for their crimes. But even those who find punishment's primary justification in its consequential efficiency in preventing crime often also insist that our pursuit of that goal must be constrained by the demands of justice or fairness: we should punish only those who deserve it, in virtue of their culpable commission of a crime (and we should not punish them more harshly than they deserve), not because it would be consequentially

---

[3] One qualification to this claim should be noted – that some of the mechanisms through which the criminal law can achieve good or avert harm might depend not on the reality of ultimate responsibility but on people's belief in it. For instance, insofar as the criminal law can retain the perceived legitimacy on which it depends for its efficacy only if people believe that criminal punishment is reserved for those who deserve it, and if people also believe that criminal punishment is deserved only by those who are 'ultimately' responsible, the general loss of any belief in ultimate responsibility would undermine the criminal law's pragmatic efficacy. Similarly, if one of the benefits of criminal punishment is the satisfaction of people's demand for retribution (see, e.g., Honderich 2005: ch. 3), and if that demand presupposes a belief in ultimate responsibility, the loss of that belief would negate that (dubious) benefit.

disadvantageous to punish those who do not deserve it, but because it would be unjust or unfair to do so.[4] For most penal theorists, and also I suspect for most citizens and policymakers, it therefore matters that punishment be just and fair, even if it also matters that it be efficacious. But if punishment can be fair only when it is imposed on those who have the kind of 'ultimate' responsibility that Strawson denies, our abandonment of any belief in such responsibility would undermine our belief in the legitimacy of criminal punishment – indeed, in the legitimacy of the enterprise of criminal law. If no one is ultimately responsible, then punishment can never be fair or just; nor, therefore, can the laws which define offences to which punishments will be attached be fair or just. We would need either to abandon the whole enterprise of criminal law, or at least to radically revise our conception of its proper aims and justification – which would be likely to involve radically revising the practice itself.

There are (at least) two ways to avoid this conclusion. One would be to argue that we (some or most of us, some or most of the time) do have the kind of ultimate responsibility that Strawson denies – which would involve finding counters to his powerful argument that such responsibility is incoherent. The other would be to argue that the denial of such ultimate responsibility does not have the devastating implications that Strawson suggests: not by denying (as a consequentialist might deny) that we need worry about the justice or fairness of punishment and of the systems of criminal justice of which it is part, but by arguing that the non-pragmatic justice of punishment does not depend on such 'ultimate' responsibility. The first response is well worth exploring, but I will not explore it here, since I will argue that a persuasive version of the second response can be provided – which undercuts the need to explore the first response. The kind of responsibility that we must have, if the enterprise of criminal law and punishment is to be consistent with the demands of justice is, I will argue, something much more modest, much less metaphysically ambitious, than the 'ultimate' responsibility that Strawson so persuasively denies.

If we are to get clear about the kind of responsibility that is relevant to criminal law, we must first get clear about the criminal law itself – about the kind of enterprise that it is, about the aims that it should serve, about the principles that should structure it, and about the conditions given which its demands on us and treatment of us can count as just. That would

---

[4] The classic modern formulation of such a view is found in Hart (1968).

be a task for a whole book – or more. All that I can do here is sketch the bare outlines of a conception of criminal law that I have developed more fully elsewhere (see Duff 2007, 2010a) to bring out the features that bear most directly on the issue at hand; that will be the task of the following section.

2. The criminal law: calling to account for public wrongs

What marks out the criminal law as a distinctive mode of legal regulation; what distinguishes it from other kinds of law? A plausible answer to this question is that, whilst the criminal law can be, and is, used to serve a variety of ends, its two central and distinctive functions are its focus on public wrongdoing, and its provision of a particular kind of formal response to such wrongdoing.

That the criminal law is concerned with wrongdoing is a familiar theme: many theorists and many practitioners would accept, for instance, that conduct should be criminalized only if it is morally wrongful – that it is unjust or illegitimate to count a person guilty of a criminal offence if we cannot argue that the conduct constituting that offence is morally wrongful in some relevant way. In two ways, however, the view offered here goes beyond that generally agreed constraint on the scope of the criminal law.

First, it claims not merely that (as many would agree) wrongdoing is a necessary condition of criminal liability, but that the criminal law is focused on wrongdoing as its primary object. It might be argued, for instance, that civil liability to pay the costs of harm that I caused ought to depend on wrongdoing, in that I should incur such liability only if I caused the harm through my own culpable conduct. The focus of such civil liability, however, is on who should have to pay the costs of the harm, and any requirement of wrongdoing figures only as a condition given which it is legitimate to require the harm-causer to bear that cost. By contrast, the criminal law makes wrongdoing salient: its definitions of offences are definitions of certain kinds of moral wrong that must be marked, avoided, and condemned as wrongs; and (as we will see shortly) it provides for an appropriate formal response to such wrongs, a response that is again focused on their wrongful character.

Second, although some theorists (so-called 'legal moralists') argue that the criminal law is in principle concerned with any and every kind of moral

182

wrongdoing,[5] that is not the view offered here. Many kinds of wrongdoing should count as 'private', in the sense that they are not even in principle the business of the criminal law: whilst we have very good reason to criminalize attacks on persons and their property, and conduct that culpably endangers others, we have no reason at all to criminalize such undoubted (and often serious) wrongs as marital infidelity, the betrayal of a friend, the callous disregard of a colleague's unhappiness, and so on. This point is sometimes expressed by saying that the criminal law is concerned only with 'public' as opposed to 'private' wrongs (Blackstone 1765–9: IV.1), which raises the question of what should count as a 'public' wrong; all I can say here (as a gesture towards the further work that needs to be done) is that a wrong counts as public, in this context, if and because it properly concerns 'the public' – in so far as it properly concerns all citizens in virtue simply of their membership of the polity whose law it is.[6]

The first distinguishing feature of criminal law, then, is its concern with public wrongs: in its 'special part', where particular offences are defined, it marks out that range of wrongs which (it is claimed) require such formal public recognition. Its second distinguishing feature is connected to the first, since it involves the provision of an appropriate response to the (actual or suspected) commission of such wrongs; we could indeed say that it completes the first feature, because one significant reason for defining such a range of public wrongs is to provide for an appropriate, formal public response to them. Theorists' accounts of that response typically focus on criminal punishment, which is indeed a distinguishing feature of criminal law. Although civil courts can award 'punitive damages' that far exceed the amount that is required as compensation for the harm caused, and although criminal courts can make compensation orders that require convicted offenders to pay compensation to their victims, it is still roughly

---

[5] See, e.g., Moore (1997: chs 1, 16, 18). As Moore makes clear, he is not committed to saying that we should actually criminalize every kind of moral wrongdoing, since there are strong reasons of both principle and practice to leave many wrongs outside the criminal law's reach. But he is committed to the view that we have good (albeit not conclusive) reason to criminalize any and every kind of moral wrongdoing – which is what I deny here.

[6] It might seem that whilst such a claim is plausible in relation to so-called '*mala in se*', those crimes (such as murder, rape, other attacks on the person, theft) that form the salient core of the traditional criminal law and involve conduct that is clearly pre-legally wrongful, it cannot be sustained in relation to the much larger class of so-called '*mala prohibita*' – offences which consist solely in the breach of a pre-existing legal regulation. For a response to this worry, see Duff (2007: chs 4.4, 7.3); Husak (2007: 103–19).

true that whereas the typical outcome of a successful civil suit is that the defendant is required to pay damages, or to make good on his prior commitments, the typical outcome of a successful criminal prosecution is that the defendant is punished. Punishments and civil damages can both be burdensome, and if the punishment consists in a fine the material burden can be of the same kind. But, first, the severity of a legitimate punishment must be proportionate to the seriousness of the past offence, whereas the amount of damages awarded depends on what it costs to repair the harm. Second, punishment is intended to be burdensome, whereas civil damages are not: any burden that an order for civil damages imposes on the defendant is a side-effect of the order's aim, whereas the imposition of an appropriate burden is central to the very point of punishment. Third, intrinsic to punishment (and to the criminal conviction that precedes and justifies it) is formal censure. The conviction does not simply record that the defendant committed what the law defines as an offence and is therefore eligible for punishment; it condemns him as a wrongdoer. His punishment also has an expressive dimension of censure or condemnation. That is the crucial difference between a penal fine and a fee: the fine does not constitute a charge for what the person has done (even if that is how many people might see some fines, as a price worth paying); it rather expresses a formal condemnation of what he has done.[7]

Now punishment (especially the obviously onerous punishments, such as imprisonment, that characterize our own penal systems) is one central feature of criminal law, which raises difficult normative questions about its legitimacy; but theorists should also pay more attention than they usually do to the formal process that precedes and legitimates punishment – in particular to the criminal trial. We might be tempted to see the trial simply in instrumental terms, as the process that links crime to punishment: the point of the trial is to identify as accurately as possible those who are eligible for criminal punishment because they have committed a criminal offence. Those who take such a view must of course admit that our existing trial processes are not well adapted to the pursuit of truth; to which they will reply that this partly reflects procedural inadequacies that should be remedied, but also partly reflects the influence of normative side-constraints on the means we may properly use to establish the truth – constraints reflected in the procedural rights that defendants are recognized

---

[7] The significance of the expressive (or communicative) dimension of punishment was brought back into the contemporary debate by Feinberg (1970).

to have. This is, however, an unduly narrow view of the criminal trial: though I cannot argue the issue here, I suggest that we should instead see the trial as a process through which someone accused of committing a public wrong is called to answer publicly to that accusation – called by the members of the polity (his fellow citizens), through the criminal court (see Duff *et al.* 2007).

More precisely, he is summoned first to answer *to* the charge that he faces – an answer that, given the presumption of innocence, need involve nothing more than a formal plea of 'Not Guilty', which requires the prosecution to adduce persuasive evidence that he committed the offence, failing which he must be acquitted. If the prosecution does discharge that probative task, the defendant is then summoned to answer *for* the offence that he has been proved to have committed. His answer might be that he has a defence – that his commission of the offence was either justified (for instance that he wounded another in self-defence) or excused (for instance that he committed perjury under a threat of serious violence that even a reasonable person might well not have resisted). If he offers evidence to support such a defence, and the prosecution fails to disprove it, he must be acquitted, since he has offered an unrebutted exculpatory answer for his commission of the offence. But if he offers no defence, or if the prosecution disproves his defence, he must be convicted.[8] A conviction does not just record a finding that he did commit the offence, and had no defence for doing so: it expresses a formal, public condemnation of his action as a wrong for which he is now formally censured by his fellow citizens – by the whole polity. He is then expected (normatively if often not predictively) to accept that condemnation as justified, and to take it to heart – indeed, to make it his own.

The trial is thus a process in which the defendant is expected to take part. It is not simply a process to which he is subjected as an essentially passive object of investigation or decision, although that is all too often how trials actually operate;[9] it is a process that is addressed to him, and that seeks a

[8] He is also convicted, of course, if he pleads 'Guilty': although such a plea is in fact often nothing more than a move in a choreographed game of plea-bargaining aimed, from the defendant's point of view, at securing a lighter sentence than he might have received had he gone to a full trial and been convicted, and from the prosecutor's point of view at avoiding the costs and uncertainties of a full trial, it is formally an answer to the charge that also constitutes answering for the commission of the offence; I answer for my wrongdoing by confessing it.

[9] It should be evident that what is offered here is not a description of the kinds of trial that

response from him, as a responsible citizen.[10] There are two dimensions to the responsibility that is involved here.

First, it matters whether the defendant was a responsible agent at the time of the alleged offence. If he was not responsible, if for instance he was then suffering from some species of responsibility-negating mental disorder, he cannot now be held responsible for his conduct; he must be acquitted on grounds of insanity.[11] Second, however, it also matters whether he is a responsible agent at the time of his trial – whether he is 'fit to plead'.[12] If he is now so disordered that he cannot understand the charge that he faces, or the trial process that deals with the charge; if he cannot answer the charge himself or instruct counsel about how to answer it; if he would be an uncomprehending observer (or object) of the trial rather than one who could participate in it: he cannot properly be tried. For the trial calls him to answer to the charge, and to answer for his commission of the offence if it is proved: if he cannot answer, it becomes a travesty of justice to call him to answer.

If we are to get clear about the implications for criminal law and criminal responsibility of a denial that any of us is ever 'truly' or 'ultimately' responsible; about whether that denial must be a denial that criminal punishment 'can be fair without having any sort of pragmatic justification', because it amounts to a denial that 'it could be just to punish some of us with (eternal) torment in hell and reward others with (eternal) bliss in heaven' (Strawson, this volume: 128): we must ask what kind of responsibility this is that the criminal law requires. It is to that task that I now turn.

---

typically take place (or, given the prevalence of bargained guilty pleas, typically do not take place) in our existing courts, but a normative account of what the trial ought to be. That normative account, however, draws on central features of our existing trial process: it is an account of what trials ought to be in terms of the values that can be discerned within our existing institutions.

[10] The defendant is of course not always a citizen, but may be a temporary resident in or visitor to the polity. Visitors are also bound by the criminal law, and called to answer in the criminal courts – as guests who are expected to respect their hosts' practices.

[11] Insanity figures formally as a defence at the criminal trial (see Note 1 above), but it differs from other kinds of defence, such as self-defence or duress, in that to plead insanity is not to answer for one's commission of the crime – it is rather to argue that one cannot be called to answer for it, because it was not the action of a responsible agent.

[12] See Robinson (1984, vol. II: 501–8); Sprack (2006: 287–8); Duff (2007: 179–81).

3. Criminal responsibility is not 'ultimate' responsibility

In defining something, call it Φ-ing, as a crime, the criminal law declares two things. It declares that we, the citizens to whom the law is addressed, have good reason not to Φ: that reason is not that the criminal law defines Φ-ing as a crime (for that would be circular); if the law is to claim our respect, it must be a moral reason that exists prior to this particular criminal law. That reason might be that Φ-ing is morally wrongful, in a way that properly concerns the whole polity, prior to and independently of any legal regulation of Φ-ing – as with so-called *mala in se*; or it might be that Φ-ing is wrong, as a breach of one's civic duties, because it breaches a legitimate legal regulation, which is itself justified as serving some aspect of the common good. In either case, the law directs our attention towards moral reasons for action that – it claims – apply to us and are binding on us. Second, however, in defining Φ-ing as a crime, the criminal law declares that it is a wrong for which we will be called to answer at a criminal trial: for that is the prescribed response to what the law defines as criminal public wrongs. In both these ways, the criminal law addresses us as agents who are in an appropriate sense 'reasons-responsive': as agents who might not in fact always and reliably respond to relevant reasons, but who are capable of doing so.[13]

Thus, first, in defining Φ-ing as a crime the law assumes that those to whom it speaks are capable of grasping and responding to, of being motivated by, the reasons not to Φ on which the law's definition of it as a crime depends. The criminal law is not binding on, and is not addressed to, children who are too young to comprehend it, or those whose serious disorder or disability cuts them off from a practically rational grasp of such reasons; it addresses, and claims to bind, only those who are within the realm or reach of reasons – of those particular kinds of practical reason in which the law deals. If those reasons are moral reasons, as I have claimed they are, to do with the moral wrongfulness of the conduct defined as criminal, the criminal law is addressed only to those who are capable of grasping and of being moved by such reasons. Thus if there is such a person as the 'partial psychopath', who is fully capable of grasping and responding to prudential reasons concerning his own self-interest, but who is incapable

---

[13] 'Reasons-responsiveness' accounts of responsibility have become popular (and increasingly complicated) in recent years: for various versions, see, e.g., Wolf (1987), Wallace (1994), Fischer and Ravizza (1998), Morse (1998).

of understanding moral reasons grounded in the interests of others or in anything other than self-interest, he is not bound by the criminal law, even though he might well be deterred by the threat of punishment – which is not to say that we must leave him free to prey on others, but that we cannot bring him within reach of the criminal law.[14]

Second, in providing for those accused of committing criminal wrongs to be brought to trial, the law assumes that those who are thus summoned are capable of answering for their actions. To be thus capable, they must be able to understand the criminal charge for what it is, as a charge of wrongdoing: they must be able to understand that they are accused of acting in a way that they had, according to the criminal law, conclusively good moral reason not to act. They must also be able to understand what it is to be called to public account in this way – to understand that they are called to answer in this forum to the polity to which they belong as citizens. And they must be capable of answering – of denying the charge against them, or of explaining their own past conduct, in the first person, in relation to that charge. If at the time of their trial they lack such capacities (and even if they were responsible agents at the time of the offence), the criminal law cannot bind them to be tried; it cannot require them to do what they are quite unable to do. That is not to say that we should then do nothing about the past crime that they allegedly committed: only that we cannot summon them to a criminal trial.[15]

Our question now is this. If a person is to be reason-responsive in these ways, must he have the kind of 'ultimate' responsibility that Strawson denies: must he be 'truly responsible for the way [he is]' (Strawson, this volume: 132)? Whether and how he responds to reasons that are put to him will indeed depend not just on what those reasons are, and how (in what tones, in what context) they are put to him, but on the way he is; and if we

---

[14] On the 'partial psychopath', see Cleckley (1964: 195–234). It is controversial whether a total inability to grasp moral or other non-self-interested values is compatible with a competent grasp of prudential values, but even if the partial psychopath is fictional, he can illustrate the point here. Note too that a more firmly consequentialist view of criminal law might produce a different conclusion: if such a person is rationally deterrable he is within the reach of a deterrent-based criminal law.

[15] Under English law, in such cases an inquiry is held, and if the jury is 'satisfied . . . that [the defendant] did the act or made the omission charged against him as the offence', a 'finding' is made to that effect, and the court then decides on a suitable disposal (Criminal Procedure (Insanity) Act 1964, ss. 4A-5). But he is not convicted, however certainly it is proved that he committed the offence and was sane at the time.

deny 'ultimate responsibility' we deny that he is ('ultimately') responsible for the way he is. But that he is not moved by certain reasons, or even that given the way he is he *will* not be moved by them, does not entail that he is not 'reason-responsive' in the (relevant) sense that he has the rational and emotional capacities given which he could be or have been moved; nor does it entail that he lacks the capacity to answer for his actions, although he might refuse to do so. To say that he is reason-responsive is to say that he could respond: that he could be moved, or could have been moved, by the reasons that did not in fact move him; that he could answer for his actions by explaining why he acted thus despite what we see as conclusive reasons against it. Now 'could' and 'can' are notoriously problematic in philosophy: one of the ongoing debates about free will concerns the possibility of analysing such locutions as 'he could have done otherwise' (locutions whose truth seems essential to responsible agency) in terms that render them compatible with the determinist thesis that in some sense, given the existing state of the world and the laws of nature, it was impossible that he would do otherwise than he did.[16] But if we think about the particular context of the criminal trial as I have sketched it here, we can see that the truth of 'he could (have) respond(ed)' in the sense that is relevant to this context, does not depend on his possession of the kind of 'ultimate' responsibility, the 'ultimate' freedom from causal determinism, that Strawson denies.

We might be confident in advance, with good reason, that a particular agent will not respond in the way that we seek: that he will, given the chance, commit a particular kind of crime; that he will not, when summoned to trial, admit his guilt or answer for what he has done in the way that he should; that he will not be persuaded, by the trial or by his punishment, to reform his ways. Analogously, we might be confident in advance, with good reason, that a moral wrongdoer will not be persuaded by anything we say to recognize his wrongdoing for what it is, to repent it, or to change his ways. If asked why we think he will be unmoved, we might talk about the entrenched aspects of his character that underpin his wrongful conduct and his resistance to our persuasive endeavours. If then asked whether we suppose that he is or was responsible for those aspects, for 'the way he is', we might (if asked outside the particular context of

---

[16] The classic discussion of the nuances of 'can' is Austin (1961). For a helpful survey of recent debates, see McKenna (2009), especially ss. 3.3, 5.1.5, and the Supplement on 'Compatibilism: State of the Art'.

philosophical discussion) be puzzled by the question, but might ask in reply whether his upbringing was such as to deprive him of exposure to the kinds of moral influence that we take to contribute to moral education, or whether he suffered some congenital incapacity that rendered him wholly insusceptible to such influences. If he did suffer some such radical malformation in upbringing or in heredity, so that he was and is cut off from any understanding of the values to which we would appeal in judging his actions or in calling him to account for them, we must of course agree that he was not and is not a responsible agent whom we can hold to account: but neither our well-grounded certainty that he will not respond appropriately, nor a Strawsonian denial of his 'ultimate' responsibility, imply that he suffers any such radical, responsibility-negating, malformation; they do not imply that he is not in the relevant sense reason-responsive.

To say that he is reason-responsive is to say, roughly, that there is an intelligible route from where he is now, in terms of his conceptions of value and of the reasons that bear on his actions, to where we think he should be and where we are trying to persuade him to be: for instance from his willing or enthusiastic commission of the crime, to a repentant recognition of the wrong that he has done. To make the idea of 'an intelligible route' slightly clearer, we could say that there is such a route so long as we could tell a coherent narrative of his persuasion: a narrative that might well be fictional, if he remains unpersuaded, but that we could imagine being true. If the person concerned is so seriously disordered as to be beyond the reach of reason, or if he is a psychopath, no such narrative is possible – no such narrative could make sense; for to be thus disordered, or psychopathic, is precisely not to be within the imaginable reach of rational moral persuasion. But absent such disorder or incapacity,[17] we can treat the person as in principle persuadable, which is what matters for responsibility; he is in principle persuadable even if we are certain that he will not in fact be persuaded.

But why should we try to persuade him, if we are sure that the attempt is doomed? Can we even be said to be *trying* to persuade him if we are sure of failure? Even if we can, won't our reasons for the attempt now have to do not with him, but with our desire to send some message to others, or to express our own feelings – in which case are we not using him merely as

---

[17] Or absent the kind of cultural gulf that might preclude mutual understanding.

a means to our own ends?[18] We certainly have reasons independent of the offender for trying to call him to account: we owe it, we might say, to the victim of his wrongdoing, and to the values he has flouted, to take that wrongdoing seriously – and calling him to account is at least a way of taking it seriously. But we also owe this to him – and that is why this is not just one appropriate way to respond to his wrongdoing, but presumptively the appropriate way to respond to serious public wrongdoing. We owe this to him because we owe it to him to treat him, as far as this is possible, as a fellow member of the normative community in which he lives (and commits his wrongs), and not to treat him as someone who is beyond moral and social redemption. It is an important aspect of our conception of ourselves and of each other as moral and social agents that we are redeemable – that we are not wholly beyond the reach of moral reason (of the good) however firmly or stubbornly we might be committed to wrongdoing. This is what makes sense of the continued effort to persuade the wrongdoer, even when we have good reason to be empirically confident that the attempt will fail. We can also see why this is important if we ask what the alternative is. It is not to ignore the wrongs that people commit, since that would be neither practically prudent nor morally appropriate: it is rather to see those who commit them not as fellow members of a normative community with whom we must engage as responsible agents, but as dangerous objects whom we must control, or incapacitate, or seek to modify.[19]

The key point to note here is that a denial of 'ultimate' responsibility does not undermine this conception of ourselves and others, or of what we owe to the wrongdoer: what matters is that it is possible to treat him, without deception or denials of what we know, as a participant in our moral and social practices; and that possibility depends on the way he now is, not on whether he was in some sense responsible for being that way.

However, it might be replied, that possibility also and crucially depends upon our ignorance. For given the truth of some form of determinism, if

---

[18] A charge disturbing for retributivists who argue that purely consequentialist conceptions of punishment illegitimately treat those punished 'merely as means' to the social goods that punishment can bring.

[19] Compare Strawson (1962) – the inspiration for much recent compatibilist work – on the 'objective' attitudes that we can take towards those whom we take not to be responsible, as contrasted with our reactive attitudes towards those whom we can see as fellow participants in our social and moral practices.

we possessed the omniscience of Laplace's imagined 'intelligence' to whom 'nothing would be uncertain', and to whose eyes 'the future as well as the past would be present',[20] we would be able to know with complete certainty whether or not a person would in fact be persuaded through the criminal process to answer for or repent what he had done; and if we knew that he would not in fact be moved by the reasons that we, or the law, could offer, the 'attempt' to persuade him would surely be empty. It is, therefore, only because we lack such omniscience that we can believe the attempt to be other than futile, which is hardly a sound basis on which to construct a system of criminal responsibility. Now there are some puzzles about the idea of such omniscient foresight: about what it would be to have it, and about just what would be foreseen; in particular, about the relationship between the movements of bodies and atoms that Laplace's intelligence would be able to foresee, and the human actions and responses that interest us. But the lack of such foresight is not merely a contingent feature of human life, as other kinds of even massive ignorance are ('If only we knew more about X, we could improve our lives'): it is a basic presupposition of a whole range of practices, attitudes and relationships. If human beings were immortal, or wholly immune to physical injury, human life and ethics would be transformed (though one might wonder whether these would still be *human* lives and ethics); but that does not undermine the categorical status of existing moral views about how we should treat each other. Likewise, the radical and perhaps unimaginable difference that such Laplacian foresight would make to human practices does not undermine the legitimacy of those human practices as practices in which we, as fallible and very far from omniscient beings, can live.

But, it might now be objected, even if it can plausibly be argued that our practices of holding ourselves and each other responsible, of calling each other to account for the wrongs that we do, of blaming and censuring each other for those wrongs, do not depend for their intelligibility or legitimacy on the kind of 'ultimate' responsibility that Strawson denies, this does not yet justify the practice of punishment, if that practice is understood

---

[20] 'Given for one instant an intelligence which could comprehend all the forces by which nature is animated and the respective situation of the beings who compose it, an intelligence sufficiently vast to submit these data to analysis, it would embrace in the same formula the movements of the greatest bodies of the universe and those of the lightest atom; for it, nothing would be uncertain, and the future, as the past, would be present to its eyes': Laplace (1902 [1814]: 4).

in retributive terms as the imposition of a burden that the wrongdoer deserves. For surely punishment, as thus understood, aspires to be a human form of the '(eternal) torment in hell' to which a vengeful Christian God was supposed to condemn sinners (Strawson, this volume: 128) and surely such penal burdens are unjust and unfair if the offender was not ultimately responsible for being or becoming the person who would commit such a wrong. Now retributive punishment is indeed sometimes portrayed, both by its advocates and by its critics, in this way: the court is to pass a definitive judgement on the defendant and his moral character – a judgement of the same kind, albeit less infallible and less all-encompassing, as will be passed by God; thus in so far as, for whatever reason, we cannot be confident of the accuracy or justice of that judgement, we cannot be confident of the legitimacy of the punishment to which it leads.[21] But that is not the only, or the most plausible, way to understand punishment as retribution.

We could instead explain punishment as a forceful continuation of the communicative process that, I have argued, the criminal trial involves: the central aim of punishment is to bring the offender to recognize the wrong he has done, to repent it, and to make appropriate moral reparation for it, by undergoing the burden that punishment imposes. I cannot discuss the details of such an account here, and do not pretend that it is free of problems or open to no objections.[22] The point to note here, however, is that this is one way in which we can understand punishment as retribution (as a deserved response to the offender's wrongdoing, which is appropriate to the character and implications of that wrongdoing), without requiring the punisher to claim a godlike power of judgement on the wrongdoer's ultimate, culpable responsibility for his character. Punishment is part (admittedly a forceful, coercive and potentially oppressive part) of our social and political dealings with each other as agents who are responsible in the modest sense sketched above. It matters that it be 'fair' independently of 'any sort of pragmatic justification' (Strawson, this volume: 128), in that we can legitimately punish someone only if and because he deserves the censure that punishment communicates, in virtue of his culpably

---

[21] Compare Murphy's discussions of 'character retributivism' (according to which penal desert is 'a function . . . of the ultimate state of one's character' (1995, 2000).

[22] See Duff (2001); for other accounts emphasizing the communicative dimension of punishment, which could also be said to avoid the objection noted here, see Von Hirsch (1993), Bennett (2008).

responsible commission of a wrong: but for it to be thus fair, for him to be thus responsible, he need not have or have had that 'ultimate responsibility' that Strawson denies; it is enough that he is, and that he was at the time of his crime, responsible in the sense of being reason-responsive.

Our practices of criminal law and punishment do indeed depend on the assumption that those engaged in them or subjected to them are responsible agents who can justly be called to account for their actions and condemned and punished for their crimes; but they do not, I have argued so far, depend on the assumption of any metaphysically puzzling kind of 'ultimate' responsibility. There is, however, another, more subtle kind of objection to those practices that might flow from the denial of such 'ultimate' responsibility, to which we must now attend.

4. 'There but for the grace of God . . .'[23]

The argument of the previous section depended in part on the claim that when we summon an alleged offender to trial, or subject him to criminal punishment, we should not do this as if we are sitting in judgement on him from on high – as if we, or the courts that act in our name, are trying to instantiate a human version of the Day of Judgement.[24] We must collectively address the defendant-offender as an equal: as a fellow-member of a normative political community, bound and protected by its values, answerable to his fellow members as they are answerable to him. But, the new objection runs, if we take this perspective seriously, and if we also deny 'ultimate' responsibility, we will find that we cannot in good conscience punish wrongdoers: for we should be held back by some version of the thought 'there but for the grace of God go I'.

There are two ways in which that thought might impinge on us. One connects it to the idea of moral luck. If we are not 'ultimately responsible' for the way we are, it is in some sense a matter of luck that we turn out this

---

[23] I'm grateful to Matt Matravers for making me confront this objection, especially to Matravers (2010); see also Matravers (2007: ch. 4) which discusses 'reasons-responsiveness' conceptions of responsibility.

[24] The 'we' here is important: in what aspires to be a liberal democracy the criminal law and its institutions must be seen not as something imposed on us as subjects by a sovereign ruler, but as formal expressions or instantiations of our own values, speaking to us (whether law-abiding or offending) in what we should be able to recognize as our own collective voice.

194

way rather than that – that I become someone with dispositions and attitudes that incline me to crime, for instance, or someone without such inclinations; and since what we do flows from the way we are, it is therefore ultimately a matter of luck that we end up doing what we do. To some people this at once undermines the idea of moral responsibility,[25] but my concern here is with the way in which it might seem to undermine not so much the very idea of moral responsibility, but our standing to blame and punish others for their wrongdoing. Once I recognize that it is ultimately a matter of luck that Jones ended up as a criminal while I did not, should I not at least hesitate before condemning and punishing him for his crimes?

The force of this question will become clearer if we look at the second way in which I might be struck by the thought 'there but for the grace of God go I'. We do not know just what kinds of factor explain why a person ends up committing a crime (although we know that the factors are highly complex, and very different in different cases); but we must surely recognize that, given the right (or wrong) background and causal influences, we (who mercifully have not committed crimes, or at least not crimes of the kind and seriousness for which Jones is now on trial) would have committed crimes of that kind ourselves. That could have been me in the dock: not because I actually committed the same kind of crime as Jones, but luckily escaped detection; nor indeed because I have the attitudes, inclinations, motivations that led him to commit his crime and that would have led me to commit a similar crime had I been exposed to the same external pressures or temptations or opportunities as him – but luckily was not thus exposed;[26] but because I could have become the kind of person who commits such crimes. There but for the grace of God.

Some versions of this thought might become incoherent,[27] but it does seem to have some real moral force: we are ill-placed to pass condemnatory judgement on the offender if we would have done what he did had we been similarly placed.[28] So does this undermine the claim that criminal condemnation and punishment can be fair and just – not because those who

---

[25] For a classic discussion of this idea of luck ('constitutive luck'), see Nagel (1978).

[26] These are species of what Nagel (1978) calls 'resultant' and 'circumstantial' moral luck.

[27] For instance, insofar as the relevant factors include, as they must, genetic features, problems about identity arise: a person born with a genetic makeup other than mine would not be me, so I cannot ask how *I* would have turned out had I been born with that makeup.

[28] I am indebted to Dennis Klimchuck (2010) for exploring this kind of barrier to condemnation; he talks of the 'moral humility' that should be induced by a recognition that we cannot be sure that we would not have acted as the agent did if placed in his situation.

face condemnation and punishment are not ('ultimately') responsible for their actions, but because we who would condemn them lack the standing to do so?

The basis of an answer to this objection was laid in the previous section, in the argument that we should not see the criminal process of trial and punishment as a matter of sitting in would-be godlike judgement on the offender who appears before (and beneath) us. It is no doubt true, and salutary to be reminded of its truth, that we too easily fall into such a view of offenders: that in penal rhetoric and policy we are too prone to distinguish 'them', the dangerous criminals, from a law-abiding 'us' who must be protected against them; that in our penal systems we too often treat offenders not as fellow members of a normative community, but as outsiders; that we think, talk and act as if 'we' would never do what they have done.[29] But, first, we should be able to see the moral deficiencies in that kind of view without appealing to a denial of ultimate responsibility: what is required is a simple (albeit demanding) recognition of civic fellowship among citizens, and an acceptance that the commission of crime should not deprive offenders of civic standing. We, and the courts and penal officials who act in our name, should address offenders as fellows: as fellow citizens who have done wrong, and who should recognize the wrongs they have done, but to whom we still owe the respect and concern that membership of the polity entails.[30] This is consistent with the practices of trials and punishments as sketched in the previous section; indeed such practices are precisely appropriate for those whom we recognize as fellow members of the polity, since they address defendants-offenders as responsible citizens who are both bound and protected by the polity's values, and who can be expected to participate in the polity's civic life.

Second, a recognition that we could have committed (that if similarly placed we might well have committed) the crime of which this person is accused should not induce an unwillingness to criticize, to censure, even to punish him. For we must still recognize and respond to the wrong as a wrong, and to its agent as a fellow member of the polity, which is what we do in calling him to account for it and in holding him responsible for it.

---

[29] Not that we can really believe that we never do or would commit what the law defines as a crime; but we tend to think and talk as if we can distinguish 'real' crimes, which 'they' commit, from technical offences of the kind that 'we' might commit.

[30] Compare Dworkin (1985: ch. 8) on the 'equal concern and respect' due between citizens of a liberal polity.

What that recognition should induce, and can reinforce, is a sense of fellowship: a readiness to help the offender face up to what he has done, and to maintain or restore our civic relationships with him; a proper acceptance of the possibility that we may be called to answer for wrongs that we commit – indeed, a readiness to answer for them; and perhaps above all, a readiness to answer to him for our conduct towards him, including such moral deficiencies as our collective treatment of him might have displayed. What it should induce, that is, is not a reluctance to call to account, or to censure, or to punish, but a practical recognition of the reciprocity of responsibility – of the way in which we must be ready and able to answer to those whom we call to answer to us.[31]

Important though the 'there but for the grace of God' reminder is, it need not and should not lead us to think that we lack the standing to call to account and to censure through the formal verdict of a criminal trial, and indeed to punish, those who commit criminal wrongs. So long as they are responsible for their deeds in the modest, 'reason-responsive' sense that I sketched in Section 2; so long as they are able to answer for themselves and for their actions as the actions of reason-responsive agents; and, crucially, so long as we ourselves are ready and willing to answer for our own misdeeds, and to answer to those whom we call to answer to us: we can still hold them responsible for what they do, and leave the question of 'ultimate' responsibility where it properly belongs – in the hands of God if we believe in God, but certainly outside the courtroom.

References

Austin, J.L. (1961) Ifs and cans. In *Philosophical Papers*, Oxford: Oxford University Press, pp. 205–32.

Bennett, C. (2008) *The Apology Ritual: A Philosophical Theory of Punishment*, Cambridge: Cambridge University Press.

Blackstone, Sir William (1765–9) *Commentaries on the Laws of England*, Oxford: Clarendon; available at www.yale.edu/lawweb/avalon/blackstone/blacksto. htm

Cleckley, H. (1964) *The Mask of Sanity*, St Louis, MO: Mosby.

---

[31] This point is particularly important when we ask about the justice of punishing those offenders (in societies like our own, those many offenders) who have suffered various kinds of systemic injustice: see Duff (2010b).

Duff, R.A. (2001) *Punishment, Communication and Community*, New York: Oxford University Press.

Duff, R.A. (2007) *Answering for Crime: Responsibility and Liability in Criminal Law*, Oxford: Hart.

Duff, R.A. (2010a) Towards a theory of criminal law? *Proceedings of the Aristotelian Society* (Supp. Vol.), 84: 1–28.

Duff, R.A. (2010b) Blame, moral standing and the legitimacy of the criminal trial, *Ratio*, 23: 123–40.

Duff, R.A., Farmer, L., Marshall, S.E. and Tadros, V. (2007) *The Trial on Trial (3): Towards a Normative Theory of the Criminal Trial*, Oxford: Hart.

Dworkin, R.M. (1985) *A Matter of Principle*, Cambridge, MA: Harvard University Press.

Feinberg, J. (1970) The expressive function of punishment. In *Doing and Deserving*, Princeton, NJ: Princeton University Press, pp. 95–118.

Fischer, J.M. (2006) The cards that are dealt you, *Journal of Ethics*, 10: 107–29.

Fischer, J.M. and Ravizza, M. (1998) *Responsibility and Control*, Cambridge: Cambridge University Press.

Hart, H.L.A. (1968) Prolegomenon to the principles of punishment. In *Punishment and Responsibility*, Oxford: Oxford University Press, pp. 1–27.

Honderich, T. (2005) *Punishment: The Supposed Justifications Revisited*, London: Pluto.

Husak, D. (2007) *Overcriminalization: The Limits of the Criminal Law*, Oxford: Oxford University Press.

Klimchuck, D. (2010) 'Excuses and excusing conditions' (unpublished).

Laplace, Marquis de (1902 [1814]) *Essai Philosophique sur les Probabilités*, trans. Truscott and Emory as *A Philosophical Essay on Probabilities*, New York: Wiley.

Matravers, M. (2007) *Responsibility and Justice*, Cambridge: Polity Press.

Matravers, M. (2010) 'Dignity without freedom' (unpublished).

McKenna, M. (2009) Compatibilism. In E. Zalta (ed.) *The Stanford Encyclopedia of Philosophy* (http://plato.stanford.edu/archives/win2009/entries/compatibilism).

Moore, M.S. (1997) *Placing Blame: A Theory of Criminal Law*, Oxford: Oxford University Press.

Morse, S.J. (1998) Excusing and the new excuse defenses: a legal and conceptual review, *Crime and Justice: An Annual Review of Research*, 23: 329–406.

Murphy, J.G. (1995) Legal moralism and liberalism, *Arizona Law Review*, 37: 73–93.

Murphy, J.G. (2000) Moral epistemology, the retributive emotions, and the 'clumsy moral philosophy' of Jesus Christ. In S. Bandes (ed.) *Law and Emotion*, New York: New York University Press, pp. 149–67.

Nagel, T. (1978) Moral luck. In *Mortal Questions*, Cambridge: Cambridge University Press, pp. 24–38.

Robinson, P.H. (1984) *Criminal Law Defenses*, St Paul, MN: West Group.

Simester, A.P., Spencer, J.R., Sullivan, G.R. and Virgo, G. (2010) *Simester and Sullivan's Criminal Law: Theory and Doctrine*, Fourth edition, Oxford: Hart.

Sprack, J. (2006) *A Practical Approach to Criminal Procedure*, Eleventh edition, Oxford: Oxford University Press.

Strawson, G. (1994) The impossibility of moral responsibility, *Philosophical Studies,* 75: 5–24.

Strawson, P.F. (1962) Freedom and resentment, *Proceedings of the British Academy,* 48: 1–25.

Von Hirsch, A. (1993) *Censure and Sanctions,* Oxford: Oxford University Press.

Wallace, R.J. (1994) *Responsibility and the Moral Sentiments,* Cambridge, MA: Harvard University Press.

Wolf, S. (1987) Sanity and the metaphysics of responsibility. In F. Schoeman (ed.) *Responsibility, Character, and the Emotions,* Cambridge: Cambridge University Press, pp. 46–62.

Guide to background reading

Mind and body

An issue lying behind many of the discussions of free will in this volume and elsewhere, and explicitly discussed in Howard Robinson's chapter, is the issue of whether conscious events (such as being in pain) are just brain events, or whether they are events of a different kind caused by brain events. The former view is physicalism, the latter is dualism. There are two kinds of dualists – property dualists who hold that the human being of whom the conscious events are properties is a physical object (that is, the human is the same thing as his or her body); and substance dualists who hold that humans consist of two parts, body and soul, and that conscious events are properties of the soul. The vast majority of philosophers and scientists are either physicalists or property dualists; but Howard Robinson defends the least popular position, substance dualism. For an introduction to the current state of the mind/body debate, see: Ian Ravenscroft, *Philosophy of Mind: A Beginner's Guide* (Oxford University Press, 2005).

Recent neuroscience relevant to free will

For a book-length introduction to the brain mechanisms involved in intentional action, discussed in the chapters by Patrick Haggard and Tim Bayne, see: Sean A. Spence, *The Actor's Brain* (Oxford University Press, 2009).

For Libet's original experimental work, see: B. Libet, *Mind Time* (Harvard University Press, 2004).

For a strong attack on the view that consciousness affects human behaviour, see: Daniel Wegner, *The Illusion of Conscious Will* (MIT Press, 2002).

For a fuller guide to the latest neuroscientific results, see: Patrick Haggard, 'Human volition: towards a neuroscience of will', *Nature Reviews Neuroscience*, 9: 934–46 (2008).

200

For detailed philosophical assessment of these results, see: Alfred R. Mele, *Effective Intentions* (Oxford University Press, 2009).

Quantum theory and the brain

For a short introduction to quantum theory, which puts most of the mathematics in an Appendix, see: John Polkinhorne, *Quantum Theory: A Very Short Introduction* (Oxford University Press, 2002). The chapter by Harald Atmanspacher and Stefan Rotter which assesses the evidence for the brain not being a deterministic system comments on the processes by which an electric impulse is transmitted from one neuron to another. Readers who need to know more about the mechanisms involved could read the article on 'Neurotransmitters and neuromodulators' in Richard L. Gregory (ed.), *The Oxford Companion to the Mind*, Second edition (Oxford University Press, 2004).

If the brain is not a deterministic system, that opens up the theoretical possibility that brain events might be influenced by conscious events. If there is this theoretical possibility, the question arises of what possible mechanism would allow this to happen. For one hypothesis about how consciousness might influence the brain, which also discusses rival hypotheses, see: J.M. Schwartz, H.P. Stapp and M. Beauregard, 'Quantum physics in neuroscience and psychology: a neurophysical model of mind-brain interaction', *Philosophical Transactions of the Royal Society* B 360(1458) 1309–27 (2005), available online at http://www-physics.lbl.gov/~stapp/stappfiles.html

Gödel's theorem

For a simple introduction to this theorem, see: T. Franzén, *Gödel's Theorem: An Incomplete Guide to its Use and Abuse* (A.K. Peters, 2005).

Moral responsibility

For an introduction to the different positions about whether physical determinism (if true) would rule out our moral responsibility for our actions; or whether, even if not all our actions are fully determined by our brain states, we could ever be morally responsible for our actions, see: Robert Kane, *A Contemporary Introduction to Free Will* (Oxford University Press, 2005).

And for a collection of articles illustrating the different positions on this issue, see: Gary Watson (ed.), *Free Will*, Second edition (Oxford University Press, 2003).

Index of names

Alvarez, Maria 150n10
Amit, D.J. 95
Aristotle xi, 102, 123
Armstrong, D.M. 52, 162n3
Arpaly, N. 33n10
Atmanspacher, Harald 4, 84, 84–101, 85, 201
Austin, J.L. 189n16
Ayer, A.J. 57n4
Azouz, R. 94

Baker, Mark 173
Banks, W.P. 26n2, 26n3, 28
Bayne, Tim 1, 2, 25–46, 33n10, 65, 160n2, 200
Beauregard, M. 201
Beck, F. 96, 97, 99
Beeson, M. 118
Benacerraf, Paul 112
Bennett, C. 193n22
Bennett, K. 39
Blackstone, Sir William 183
Boolos, George 109n3, 124n2
Brass, M. 10
Brunel, N. 95
Bryant, H.L. 93
Buridan, Jean 29

Calvin, John x
Campbell, C.A. 137
Cantor, Georg 119
Carr, E.H. 129
Chalmers, D. 162n3
Chisholm, R. 165n8, 172n11

Chomsky, Noam 173–4
Church, Alonzo 106, 107
Clarke, R. 131n4
Cleckley, H. 188n14
Clifford, W.K. x
Collins, R. 161

Darwin, Charles 135
Davidson, Donald 59, 73n9
Davies, M. 59
Davis, M. 107
Dedekind, Richard 103
Deecke, L. 15, 40
Deiber, M.P. 10
Della Sala, S. 14
Dennett, Daniel 36, 145n5, 173
Derrida, Jacques 171
Descartes, René 162, 163–4
Desmurget, M. 19
Doe, John 54
Doris, J. 133n5, 139n7
Dray, W.H. 125n3
Duff, R.A. 5, 6, 178–99, 182, 183n6, 185, 186n12, 193n22, 197n31
Dworkin, Richard 196n30

Eccles, J. 96, 99
Eimer, M. 18, 27n5, 41
Epicurus x
Euclid 103

Feferman, Solomon 4, 102–22, 109n3, 114n7, 116, 118, 123–5
Feinberg, J. 184n7

Fischer, John Martin 29n6, 131n4, 148n9, 155, 179n2, 187n13
Flanagan, O. 30
Fodor, Jerry A. 73n10, 173
Foster, John 162
Frankfurt, Harry 57n4, 141, 142–4, 146–9, 152
Franzén, T. 105n1, 201
Freeman, W. 95
French, S. 170n9
Fried, I. 19

Gaillard, Raphael 68n5
Geach, Peter 172, 174
Gerstein, G.L. 94
Gerstner, W. 92n6
Gibbs, Josiah Willard 109n2
Glimcher, P.W. 84
Gödel, Kurt 4, 102–22, 123–5
Gomes, G. 26n2, 41
Gray, C.M. 94
Gray, J. 68n5
Gregory, Richard L. 201
Gustafson, K. 90

Haggard, Patrick 1, 2, 7–24, 10, 12, 17, 18, 26n2, 27n5, 29, 39n15, 40–1, 65, 160n2, 200
Hakim, V. 95
Hallett, M. 26n3
Hameroff, S.R. 97
Harman, G. 133n5
Harnish, R.M. 120
Hart, H.L.A. 181n4
Hart, W.D. 163
Hartmann, L. 97n7
Hebb, D.O. 92n6
Heckhausen, H. 27n5
Heisenberg, Werner 85n2
Hepp, K. 97
Herbrand, Jacques 106, 107
Hilbert, David 103–4, 106

Hobbes, Thomas xii, 149n9
Hodgkin, A.L. 93
Holton, R. 33, 43
Honderich, T. 180n3
Horgan, T. 34n11
Hume, David xii, 149n9, 171
Husak, D. 183n6
Huxley, T.H. x
Huxley, A.F. 93

Ikeda, A. 15
Isham, E.A. 26n3

Jackson, Frank xiv, 1, 2, 47–62, 59, 63, 65, 159n1
Jahnke, S. 95
James, William x, xii
Jammer, M. 85n2

Kallestrup, J. 39
Kane, Robert 130, 136, 202
Kaneko, K. 95
Kant, Immanuel 129–30, 174
Kay, K.N. 79n14
Keller, I. 27n5
Kellis, S. 80n15
Kim, J. 39
Kinsbourne, Marcel 36
Klimchuck, Dennis 195n28
Knobe, J. 139n7
Koch, C. 94
Kornhuber, H.H. 15
Korsgaard, Christine 130
Kreisel, Georg 111–12

Laming, D.R.J. 75n12
Laplace, Pierre-Simon ix, 192
Larmer, R. 161
Lau, H.C. 27n5
Levy, Neil 26n2, 29n6, 33n10
Lewis, David 171n10
Li, J. 97n7

Libet, Benjamin xiii–xiv, xv, 1, 16–18, 26–31, 33n9, 34–7, 39–42, 49, 65, 81, 160n2, 200
Lindström, Per 115
Locke, John xi
Lucas, J.R. 4, 102, 109, 112–14, 115, 117n8, 120, 123–5, 132n5
Ludlow, P. 159n1
Luther, Martin x
Lycan, W.G. 74n11
Lyngzeidetson, A.E. 119n9

Mac Lane, S. 118
MacKay, D.M. 129n2
Madell, Geoffrey 166
Magno, E. 12
Mainen, Z.F. 93
Mandelbrot, B. 94
Markram, H. 92n6
Matravers, Matt 194n23
Maxwell, James Clerk 87
McCallum, W.C. 40
McClelland, John 120
McDowell, J. 164n7
McKenna, M. 189n16
Mele, Alfred R. 26n2, 38, 40, 41, 201
Mill, John Stuart xii, 18
Miller, J. 27n4, 27n5, 40n16
Moore, M.S. 183n5
Morse, S.J. 187n13
Murphy, J.G. 193n21

Nadel, L. 26n2
Nagel, Thomas 109n3, 195n25, 195n26
Nahmias, E. 32n8, 139n7
Neher, E. 92
Neuper, C. 15
Newsome, W.T. 94
Newton, Isaac ix, 87, 174
Nichols, S. 32, 139n7
Nietzsche, F.W. 134, 171
Nussbaum, M. 130n3

Paraoanu, G.S. 97n7
Patten, John 130
Peano, Giuseppe 103, 104, 106, 110, 117
Penrose, Roger 97, 102, 114–15, 120, 123, 125, 173n12
Peters, A.K. 201
Plato 162
Pockett, S. 26n2, 26n3, 28, 31n7
Polkinghorne, John 201
Primas, H. 85
Prinz, J.J. 74n11
Purdy, S. 31n7
Putnam, Hilary 112, 162n3

Ramsey, F.P. 76n13
Ravenscroft, Ian 200
Ravizza, M. 29n6, 187n13
Reid, Thomas 165
Rigoni, D. 40
Robinson, Daniel 63n1
Robinson, Howard 2, 5, 64n2, 158–77, 159n1, 162n3, 162n5, 163n6, 173n12, 175n14, 200
Robinson, P.H. 186n12
Robinson, R. 68n5
Roediger, H.K. 26n3
Rosenthal, David 37
Roskies, Adina 28–9, 40, 43
Ross, W.D. 151n13
Rotter, Stefan 4, 84, 84–101, 201
Rumelhart, David 120
Russell, Bertrand 103, 172
Ryle, Gilbert 61, 102, 123, 162n3, 176

Sakmann, B. 92
Sartre, Jean-Paul 129, 141
Scheibe, E. 85
Schrödinger, E. 87, 89, 96
Schwartz, J.M. 97, 201
Segundo, J.P. 93
Sejnowski, T.J. 93
Shadlen, M.N. 94

Shapiro, Stewart 115
Shea, Nic 50n1
Sher, G. 33n10
Sherrington, C.S. 12
Shinkareva, S.V. 69n7
Simester, A.P. 179n1
Simons, Peter vii–xv, 1n1
Sinnott-Armstrong, W. 26n2
Skinner, B.F. 10
Smilansky, S. 138n7
Smith, A. 33n10
Smith, P. 105n1
Softky, W.R. 94
Solomon, Martin 119n9
Sompolinsky, H. 95
Soon, C.S. 41n17
Spence, Sean A. 26n3, 200
Spinoza, Benedict de x, xv
Sprack, J. 186n12
Stapledon, Olaf 176n15
Stapp, H.P. 97, 161, 201
Steward, Helen 1, 5, 6, 141–57, 148n9, 150n11
Strahm, T. 118
Strawson, Galen 1, 4–5, 6, 25, 58, 126–40, 129n2, 131n4, 151, 172, 174–5, 179, 181, 186, 188, 191n19, 192–4
Strawson, P.F. 138n7
Sumner, P. 14

Swinburne, Richard 1–6, 49, 63–83, 64n2, 66n3, 66n4, 162, 174n13

Tarski, Alfred 119
Tolhurst, D.J. 94
Trevena, J.A. 27n4, 27n5, 40n16
Tsuda, I. 95
Tuckwell, H.C. 94
Turing, Alan 103, 106–8, 116, 120
Tye, M. 36
Tzagarakis, C. 15

Van Vreeswijk, C. 95
Von Hirsch, A. 193n22
Von Neumann, John 106, 120

Wallace, R.J. 187n13
Wang, Hao 111
Watson, Gary 202
Wegner, Daniel M. 19, 26n3, 200
Wenke, D. 19–21
Whitehead, Alfred North 103
Williamson, T. 52
Wilson, T. 139n7
Wittgenstein, Ludwig 11
Wolf, S. 187n13

Zak, Paul J. 80n16
Zermelo, Ernst 103, 117, 118, 123
Zumdieck, A. 95